Business Aspects of Web Services

Christof Weinhardt • Benjamin Blau
Tobias Conte • Lilia Filipova-Neumann
Thomas Meinl • Wibke Michalk

Business Aspects of Web Services

Authors
Christof Weinhardt
Karlsruhe Institute of Technology
Institute of Information Systems
and Management
Englerstr. 14
76131 Karlsruhe
Germany
christof.weinhardt@kit.edu

Benjamin Blau
Karlsruhe Institute of Technology
Institute of Information Systems
and Management
Englerstr. 14
76131 Karlsruhe
Germany
benjamin.blau@kit.edu

Tobias Conte
Karlsruhe Institute of Technology
Institute of Information Systems
and Management
Englerstr. 14
76131 Karlsruhe
Germany
tobias.conte@kit.edu

Lilia Filipova-Neumann
Research Center for Information
Technology
Haid-und-Neu-Str. 10-14
76131 Karlsruhe
Germany
filipova-neumann@fzi.de

Thomas Meinl
Karlsruhe Institute of Technology
Institute of Information Systems
and Management
Englerstr. 14
76131 Karlsruhe
Germany
thomas.meinl@kit.edu

Wibke Michalk
Karlsruhe Institute of Technology
Institute of Information Systems
and Management
Englerstr. 14
76131 Karlsruhe
Germany
wibke.michalk@kit.edu

ISBN 978-3-642-22446-1 e-ISBN 978-3-642-22447-8
DOI 10.1007/978-3-642-22447-8
Springer Heidelberg Dordrecht London New York

Library of Congress Control Number: 2011938282

ACM Codes: K.4.4, K.6.0, H.3.5

Printed on acid-free paper

Springer is part of Springer Science+Business Media (www.springer.com)

Foreword

In the 1990s, enterprises primarily optimized internal business processes to reduce process cost and to increase productivity leveraging Enterprise Resource Planning (ERP) application systems. Then the same enterprises decided to concentrate on their core business and outsourced their "non-core" activities. As a result they reduced their level of integration.

Therefore, inter-company application software like Supply Chain Management or Electronic Data Interchange technologies crossed the bridge between enterprises and increased the level of integration, but to a significant price for the parties in the value chain.

With the Internet and related Web service technologies like REST and SOAP becoming popular, it is now possible to out-task even fine granular activities (e.g., an external credit check service) within minutes.

This effect becomes accelerated with the boom of software-as-a-service (SaaS) applications that rely very much on Web services to integrate with other SaaS applications as well as with internet-based "e-services" in general – even next generation enterprise application software like SAP Business ByDesign leverages Web services for the internal communication between functional building blocks.

In essence, Web service has become the glue holding companies in their value chain together as "service value networks." While a lot of technological-oriented books have been published on Web services, there is still a gap in research to look at the business impact of Web services as e.g., finding of the right Web service via marketplaces, aspects of composition, business models and pricing (strategies) of Web services.

I am happy to say now that with this book here, Christof Weinhardt and his colleagues have created a compelling book about this topic that is able to bridge this research gap and covers the main business aspects of Web services.

Walldorf
Germany

Dr. Wolfgang Faisst

Preface

In April 2009, one of the authors of this book, Thomas Meinl, was attending the World Wide Web conference in Madrid, Spain. There, he gave a talk, in which he presented the idea of Web service reservation via derivative mechanisms. During the following discussions, it became clear that for many researchers, practitioners and managers this topic, while it even only scratched basic economic and financial principles, without going into any mathematical details, was still completely beyond their scope. This became even more surprising when bearing in mind that, as most Web services are built in order to generate some quantitative (financial) profit, the related economic aspects should be considered from the very start. Therefore, even if economic issues were, are and should not be the main concern of Web service developers and decision makers, Thomas Meinl figured that it could prove to be advantageous if they had at least a general notion of the processes following their work.

Therefore, the idea came up to write a book that would make both these basic relationships and deeper economic coherencies accessible to a wider audience, in this case, an audience particularly belonging to what is known as the Web community. It was clear from the very beginning that such a book should, on the one hand, contain parts that are hands-on so that the economic principles and foundations presented there could be applied directly without necessarily consulting further literature. On the other hand, we deliberately aimed to also include more theoretic contents in order to also provide for a deeper mathematical background of the economic aspects to be presented. In order to account for a hands-on as well as a deeply theoretically founded perspective, the scope of this book covers established economic results on the hand and recent developments on the other hand. Yet, it is presented in a style that makes it comprehensible to the above mentioned audience. Thus, the purpose of the book is to pinpoint the relevance of economic principles in the vast field of Web services, and Web service networks, in particular, that has been considered largely from a technical perspective so far. In this vein, basic principles shall be illustrated by demonstrative examples and, where possible, extended onto Web services directly.

We would like to thank our families, for their patience, and Jack, the dog, who became our mascot while writing this book. We also thank Ralf Gerstner from Springer Publishing, who initiated and supported the idea of this book from the very beginning. Of course, we must not forget our colleagues who contributed with ideas, discussions and in other ways to many of the results presented in this book. We are deeply grateful to all of you.

Karlsruhe
Germany

Christof Weinhardt
Benjamin Blau
Tobias Conte
Lilia Filipova-Neumann
Thomas Meinl
Wibke Michalk

Contents

Chapter 1
Introduction

Since the end of the 1990s, the software industry has undergone tremendous changes. Driven by maturing Web service technologies and the wide acceptance of the service-oriented architecture paradigm, the software industry's traditional business models along with business strategies have already started to erode – with far-reaching consequences: software vendors turn into service providers. While traditional software products are installed at the customer site, including prepaid perpetual-use licences, so-called software-as-a-service (SaaS) or on-demand software is hosted and maintained by the service provider itself that offers usage- or subscription-based pricing models (Dubey and Wagle 2007; Choudhary 2007a,b; Sääksjärvi et al. 2005). Salesforce.com's Sales Cloud 2 is repeatedly referred to as a prime example for SaaS, mapping valuable customer relationship management (CRM) software into an online service infrastructure that can be accessed via Web browsers.[1] Exploiting the capabilities of service-oriented architectures, such services are made available "one-to-many," that is, a multitude of application instances can be run at once in a common environment. In the last decade, a large body of surveys and studies has been published which indicate the enormous potential of on-demand service provisioning. According to a series of Gartner studies, the worldwide market for on-demand enterprise applications increased by more than 25% from 2007 to 2008 (adding up to revenues of $6.4 billion) and is supposed to more than double until 2012, which will account for approximately 25% of the entire enterprise application market. SaaS is expected to exhibit consistent compound annual growth of roughly 20% through 2013 which is nearly five times the growth predicted for the total market (Mertz et al. 2007, 2008, 2009).

The prosperous future accredited to the SaaS market by a multitude of surveys seems to turn into reality: offerings that have been in the market for several years constantly exhibit almost skyrocketing usage figures. For instance, in 2007, the bandwidth consumed by Amazon's Web service offerings such as the Elastic

[1] http://www.salesforce.com/crm/products.jsp.

C. Weinhardt et al., *Business Aspects of Web Services*,
DOI 10.1007/978-3-642-22447-8_1, © Springer-Verlag Berlin Heidelberg 2011

Compute Cloud (EC2)[2] or the Simple Storage Service (S3)[3] for the first time exceeded the bandwidth consumed by all global Amazon shopping Web sites.[4] While salesforce.com reported roughly 30,000 paying subscribers in 2001, the subscription figures exploded to more than 2 million by the end of 2010.[5]

While the success story of on-demand software is likely to continue, a second wave of innovation has great potentials to shake the software industry's foundations once again. Exploiting the capabilities of Internet standards and interoperability, joint value creation of service providers has emerged. Open standards and service-oriented architectures constitute important building blocks for innovative Web service networks, tying together the competencies of specialized contributors. This development was first described in Coase's (Coase 1937) seminal paper on companies' sizes. Once serving the whole value chain, celebrating what has become well-known as vertical integration, companies now focus on their core competencies. If transaction costs in the open market decrease, companies will consequently downsize: vertical integration is abandoned in favor of flexible cooperation between adaptive and lean competence-orientated partners (Coase 1937; Downes and Mui 2000; van Heck and Vervest 2007). In the Web service market, a multitude of small and highly specialized providers offer modular services of almost any kind. Such market structures imply that contracting out is more efficient than relying on internal transactions.

The adaptiveness of the partners coincides with the development of software customers demanding more sophisticated as well as more specialized solutions and, at the same time, longing for more flexible service provisioning (Bovet and Martha 2000; Downes and Mui 2000; Tapscott et al. 2001; Prahalad and Ramaswamy 2004; Basole and Rouse 2008). One of the most powerful approaches to handle complexity is modularity, that is composing the ensemble from smaller subsystems that are designed independently, yet function together as a whole (Baldwin and Clark 2000). Along those lines, vendors concentrate on their core activity while leveraging knowledge and assets of complementary partners. That way, they are able to stay agile and to flexibly adapt their services to changes in the environment, be it customer-, competition-, or regulation-driven (Sambamurthy et al. 2003; van Oosterhout et al. 2006). Such joint value creation in terms of Web services is mostly coordinated by a mediating entity as present in today's leading service platforms: Salesforce.com offers its on-demand service market place AppExchange[6] and its development platform force.com,[7] Xignite operates the Splice Mashup Platform,[8]

[2]http://aws.amazon.com/ec2/.

[3]http://aws.amazon.com/s3/.

[4]http://aws.typepad.com/aws/2008/05/lots-of-bits.html, accessed on 03/21/2011.

[5]https://www.salesforce.com/company/investor/financials/.

[6]http://sites.force.com/appexchange/apex/home/.

[7]http://www.salesforce.com/platform/.

[8]http://splice.xignite.com/.

and StrikeIron has the IronCloud Web services delivery platform[9] ready, just to name a few. Recent surveys predict that such SaaS platforms will be a crucial factor for an even broader SaaS adoption (Koenig et al. 2006; Dubey et al. 2008).

However, besides the above-mentioned increase in customers' demand for complex applications and the resulting agility of service providers to tackle these requirements, other concrete economic factors are driving this second innovation wave of the software industry. In this book, it is argued that it has its seeds in the *long tail phenomenon* which was initially promoted by Anderson (2004). The original long tail story was bred by the emergence of e-business which made it possible to transfer traditional physical business to the Internet. In online stores, merchandise assortments can be displayed without the physical restrictions present in a brick-and-mortar store. What is more, Anderson (2004) sees customers freed from the tyranny of compromise-ridden mass products by the possibility of pushing the niche through information and communication technology (ICT). Altogether, since customers highly value the new possibilities of accommodating specialized demand, the large amount of small sales of specialized products has the potential to overcompensate revenues generated by selling mass products.

The long tail's striking relevance for electronic services, and Web services in particular, has been largely neglected by academic literature so far. Requirements for functional and non-functional characteristics of Web services are much more pronounced and specific than in other domains. Such specificity of requirements considerably intensifies the niche effect. For instance, several goods exhibit the so-called blockbuster characteristic: The availability of highly specialized offers does not cut back the success of products that are designed to fit the mass appeal. This is certainly not true for Web services to such an extent. If the customer can choose between a Web service that perfectly fits his needs and a Web service that is programmed to capture the mass, he will most probably go for the former – if priced appropriately. However, not only the specific requirements make the long tail phenomenon important for the Web service domain. Modular services can be combined and configured into value-added complex services which have the potential to meet virtually every conceivable customer requirement, giving rise to a new level of customization. Such complex services involve the assembly and invocation of several specialized service modules offered by a multitude of expert partners in order to accomplish a multi-step business functionality Papazoglou (2007). Recombining service modules, new functionality is created "off-the-shelf". Beyond that, such individual composition of Web services is capable of even more: Web services are invoked more often if they provide added value in a multitude of complex services. In fact, if n services are registered with a service platform, up to $\sum_{k=1}^{n} \frac{n!}{(n-k)!}$ alternative service mashups could be created, each of them potentially meeting a specialized demand. That way, service providers can exploit economies of scope (Panzar and Willig 1981): their offerings are not designed to

[9] http://www.strikeiron.com/Company/IronCloud.aspx.

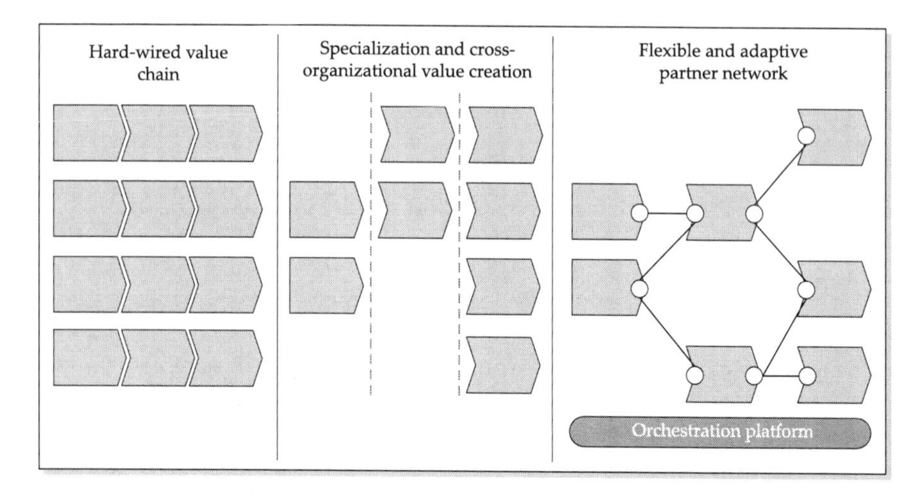

Fig. 1.1 From hard-wired value chains to adaptive service value networks. Adapted from Heuser et al. (2007)

be included in one or few complex services, but contribute to various solutions capturing diversified customer needs.

From a technical perspective, dynamic Web services are increasingly used in the context of service mashups, facilitating lightweight approaches such as RESTful architectures (Fielding 2000; Richardson and Ruby 2007) and slim messaging formats such as JSON (Crockford 2006). The service mashup platform ProgrammableWeb reported that 73% of all listed APIs exposed REST interfaces by March 2011, foretelling the trend to an internet of interoperable Web services.[10] Economically, value is created through the interplay of various distributed service providers that jointly contribute to an individualized and integrated solution. However, not only partners offering complements constitute the long valley, but also substitutive services and vendors that serve the same customer segments. Thus, service providers find themselves in the fruitful state of co-opetition, breeding both complementary opportunities and competitive threats (Brandenburger and Nalebuff 1996). While cooperation enables advanced value creation and the access to partners' assets and knowledge (Bengtsson and Kock 2000), the competitive component diminishes adverse effects of market power and spurs improvement and innovation pressure (Porter and Millar 1985; Ilinitch et al. 1996).

The above-introduced second innovation wave of the software industry (cp. Fig. 1.1), most notably the combinatorics in service mashups, can be optimally catalyzed by universally accessible service orchestration platforms – service value networks (SVNs) – which are the underlying organizational form of the coordination mechanisms presented in this book. Thereby, optimality denotes an SVN's ability to

[10]http://www.programmableweb.com/apis, accessed on 03/23/2011.

exploit the exponentially growing number of service mashups which are offered to customers (1) in an automated fashion and (2) via a permanently available channel.

This book focuses on providing comprehensive business-oriented insights into today's trends and challenges that stem from the transition to a service-lead economy. The goal is to foster a broad but at the same time in-depth understanding of current developments, opportunities and risks in the context of Web services and their business aspects.

The book investigates actual and future Web service business models and provides a framework for Web service Value Networks. An essential element of business models is to conceive the sources of profit and how it can be realized, which includes pricing first and foremost. Joint value creation by providing modular Web services in SVNs creates new opportunities for pricing but also constraints to the pricing mechanisms which can be applied. Pricing mechanism basics are introduced and transferred to this specific application area, subject to different perspectives: Strategies for platform providers are analyzed from the viewpoint of a single provider as well as pricing mechanisms in SVNs which are optimal from a network perspective. Also the extended concept of pricing Web service derivatives is shown.

The book concludes with the vision of how Web service markets in the future will possibly be structured and what further developments can be expected to happen. Having read this book, the reader will have a thorough understanding of economic and business aspects which can be applied directly to already established Web service frameworks.

This book addresses applied researchers – with respect to teaching activities and industry- or business-oriented research – as well as practitioners – CIOs, CEOs, and managers of SMEs in the service sector. The mission is to help these groups of interest to understand the organizational, economic and technical implications of dealing with Web services as the nucleus of modern business models. The book also addresses groups familiar with the technical aspects of Web services and offers a perspective onto the economic aspects which can be applied to Web services in general and Web service Value Networks specifically.

Since the book does not address the technical aspects of Web services and the respective Value Networks, prerequisites are basic mathematical calculus on a graduate student's level, which can be expected from the main target group. While some chapters will require some deeper knowledge of some mathematical areas (e.g., stochastic processes in 2.6) in order to thoroughly understand all details of the presented models, this will only be necessary for actual implementations, while the general idea should remain clear to all readers.

Although the chapters in this book build on each other by starting from the definition of Web services, followed by the introduction of general and subsequently more specific concepts of the economic environment of Web services, the chapters are not necessarily meant to be read in the same order. Each of the following chapters is based, at least partially, on work independently published by the authors elsewhere in the past few years. These pieces of work which focus on different economic aspects of Web services are organized in a common structure in this book in order to provide an overall perspective. At some places, the original publications are briefly summarized and, at other places, they are literally adopted or slightly changed to fit the purpose of this book.

The structure of this book is as follows:

After this introduction, of which the motivation bases on Blau et al. (2010), Conte (2010), the first part of this book focuses on Web service Business Models.

Chapter 2 provides a comprehensive understanding of Web services. A service typology will be given that facilitates a differentiation of Web services as opposed to Services in general and e-Services in particular. This differentiation is applied by means of service characteristics such as input and outcome transfer, existence of well-defined interfaces etc. Based on the definition of Web services, their characteristics (like intangibility, inseparability, perishability, and long-tail aspects) are derived, and the implications for pricing is analyzed. Furthermore, the chapter highlights current Web service description standards and the (non-)existing ways to formulate prices. The need for specifications that allow for flexible pricing mechanisms, which are the subject of Part II of this book, are derived and its influence on current standards is be shown. A preliminary version of the service definition has been presented in Conte (2010).

The current industry-driven trend of providing flexible e-services lays the ground for the new research area "SVNs" which is the subject of Chap. 3. When it comes to formalizing and economically analyzing such SVNs that offer joint complex services to service customers, scientific approaches are in their infancy. We intend to fill this research gap by providing a clear understanding of SVNs by defining their characteristics, their structure, and their components. Mapping these aspects into a formalized model, we intend to establish a reference point for future work in the area of SVNs. This chapter is based on and extends an article by Blau et al. (2009). Parts of Chap. 3 are also based on Conte (2010).

Chapter 4 deals with business models, which are oftentimes considered as "the most discussed and least understood aspect of the web" (Rappa 2001). The existing literature lacks contributions that particularly consider business model frameworks for networked economies, especially newly emerging loosely-coupled configurations as assumed in the service value network theory. In order to address these shortcomings, a business model framework for business value networks is introduced as a result from an extensive literature review. A state-of-the-art analysis of existing definitions of the term "business model" with emphasis on the elements that should be included is performed as well as a differentiation from related concepts is provided. For the design of a business model framework, characteristics of service value network as defined in Chap. 3 are incorporated. As a result, the partner network and the roles of these actors, their core competencies, and the role of the customer are identified as crucial elements that are to be considered in business models of companies acting in SVNs.

Part II of this book starts with a brief overview of the pricing foundations in Chap. 5. Various concepts such as fixed pricing, personalized pricing, versioning and bundling, group oriented pricing and auctions, which are known from pricing of conventional goods and services, are explained and discussed as to their application to Web services.

Chapter 6 deals with the pricing decisions of a service provider. Thereby, it takes the perspective of a single service provider in his role of an intermediary interacting with other providers and customers in a Service Value Network. The value creation in SVNs can take place in a successive manner along the value chain both in terms of the provision of the single service components as well as in terms of establishing the required service level agreements. Accordingly, providers of the single service components are required to enter into Service Level Agreements with their counterpart at the preceding and the subsequent value-added step. This chapter provides a better understanding of how prices and penalties, as part of the service level agreements, will be set when service providers compete for the provision of complementary modular services which are composed to a complex service. It takes explicitly into account that the probability of service delivery is endogenous and depends on the contracting terms. This chapter is a version of Filipova-Neumann et al. (2010) for which the original scenario with a central intermediary was modified in a way that service providers interact with each other along a value chain.

Chapter 7 deals with pricing in SVNs from an overall system perspective. Coping with the challenge of coordinating value creation in SVNs, it presents the design of a multidimensional procurement auction for trading service compositions – the Complex Service Auction – following a mechanism design approach. The mechanism implementation is incentive compatible, meaning that it results in an equilibrium in which revealing the true multidimensional type (quality of service and valuation) is a weakly dominant strategy for all service providers. This chapter introduces a mechanism that implements a novel way of distributing revenues among service providers – the power ratio. The power ratio rewards vendors for their readiness to offer services in the network, not merely for actual provision of a particular service. This approach is proven to be able to create a fair distribution of the generated value within the network and to draw in service providers and thereby foster the growth of the network. This chapter bases on previous work from Conte et al. (2009; 2010a,b), Conte (2010).

Chapter 8 focuses on the aspect of Web service reservation, that is, contracts that ensure the availability of a Web services at a certain price. The concept of reservation can be extended to the more general notion of derivatives. A derivative pricing model is deduced from Web service characteristics described in Chap. 2, leading to the concept of incomplete markets, which, from a financial mathematical point of view, is the only valid derivative pricing model for (Web) services. Further on analogies to other derivative markets (e.g., electricity) are shown. Also the structure of Web service derivative markets and the incentives to trade therein are discussed. Furthermore, the Web service derivative model is compared to an alternative approach that is heavily used in airline reservation systems, that is, revenue management. Both methods are compared and discussed regarding their requirements and their practical realization. Some of the main ideas of this chapter were already published in Meinl and Blau (2009), Meinl et al. (2010).

Finally, Chap. 9 summarizes the insights from the previous chapters and characterizes the vision of Web services. In this chapter, the authors try to give an outlook for the direction in which Web service markets will evolve in the future and to identify possible challenges. This includes the attempt to make a forecast both of possible technological attainments as well as of changes in the organizational and economic environment.

References

Anderson C (2004) The long tail. The Wired Mag 10:100–177

Baldwin C, Clark K (2000) Design rules, Vol. 1: The power of modularity. MIT, Cambridge

Basole R, Rouse W (2008) Complexity of service value networks: Conceptualization and empirical investigation. IBM Syst J 47(1):53–70

Bengtsson M, Kock S (2000) "Coopetition" in business networks – to cooperate and compete simultaneously. Ind Market Manag 29(5):411–426

Blau B, Krämer J, Conte T, van Dinther C (2009) Service value networks. In: Hofreiter B, Werthner H (eds) Proceedings of the 11th IEEE conference on commerce and enterprise computing (CEC), Vienna, pp 194–201

Blau B, Conte T, Weinhardt C (2010) Incentives in service value networks – on truthfulness, sustainability, and interoperability. In: ICIS 2010 Proceedings, Saint Louis, MO, USA, paper 8

Bovet D, Martha J (2000) Value nets: Breaking the supply chain to unlock hidden profits. Wiley, New York

Brandenburger A, Nalebuff B (1996) Co-opetition. Doubleday, New York

Choudhary V (2007a) Comparison of software quality under perpetual licensing and software as a service. J Manag Inf Syst 24(2):141–165

Choudhary V (2007b) Software as a service: Implications for investment in software development. In: Proceedings of the 40th Annual Hawaii international conference on system sciences (HICSS), Waikoloa, p 209a

Coase R (1937) The nature of the firm. Economica 4(16):386–405

Conte T (2010) Value creation through co-opetition in service networks. Dissertation, Karlsruhe Institute for Technology

Conte T, Blau B, Satzger G, van Dinther C, Weinhardt C (2009) Rewarding contribution to service network formation. In: Proceedings of the 1st INFORMS international conference on service science, Hong Kong, paper ID 225

Conte T, Blau B, Knapper R (2010a) Networked mechanism design – incentive engineering in service value networks as exemplified by the co-opetition mechanism. In: Proceedings of the 16th Americas conference on information systems (AMCIS), Lima, paper ID 383

Conte T, Blau B, Xu Y (2010b) Competition of service marketplaces – designing growth in service networks. In: Proceedings of the 18th European conference on information systems (ECIS), Pretoria, paper ID 287

Crockford D (2006) The Fat-Free Alternative to XML. in: Proceedings of XML

Downes L, Mui C (2000) Unleashing the killer App: Digital strategies for market dominance. Harvard Business School Press, Boston

Dubey A, Wagle D (2007) Delivering software as a service. Online, URL http://www.mckinsey.de/downloads/publikation/mck_on_bt/2007/mobt_12_Delivering_Software_as_a_Service.pdf, the McKinsey Quarterly, Web exclusive

Dubey A, Mohiuddin J, Baijal A, Rangaswami M (2008) Enterprise software customer survey. Sand Hill Group, McKinsey and Company, http://www.interop.com/downloads/mckinsey_interop_survey.pdf

Fielding R (2000) Architectural styles and the design of network-based software architectures. PhD thesis, University of California, Irvine

Filipova-Neumann L, Michalk W, Blau B, Weinhardt C (2010) Incentivation of service providers for efficient service composition agreements. In: Proceedings of the international conference on e-Commerce, e-Business and e-Service 2010 (ICCBS 2010), Paris, France, pp 1569–1583

van Heck E, Vervest P (2007) Smart business networks: How the network wins. Commun ACM 50(6):28–37

Heuser L, Lacher S, Perlmann S (2007) Flexible prozessgestaltung als basis innovativer geschäftsmodelle – von der service-orientierten architektur zur vision des business webs. In: Proceedings of the Wirtschaftsinformatik (WI), Karlsruhe, pp 19–28

Ilinitch A, D'Aveni R, Lewin A (1996) New organizational forms and strategies for managing in hypercompetitive environments. Organ Sci 7(3):211–220

Koenig M, Guptill B, McNee B, Cassell J (2006) Saas 2.0: Software-as-a-service as next-gen business platform. Saugatuck Strategic Research Report

Meinl T, Blau B (2009) Web service derivatives. In: Proceedings of the 18th International on World Wide Web (WWW), Madrid, pp 271–280

Meinl T, Tatsubori M, Anandasivam A (2010) Enabling cloud service reservation with derivatives and yield management. In: 12th IEEE conference on commerce and enterprise computing, E-Commerce and E-Services, Shanghai, pp 3–10

Mertz S, Eschinger C, Eid T, Pring B (2007) Dataquest insight: Saas demand set to outpace enterprise application software market growth. Gartner report

Mertz S, Eid T, Eschinger C, Swinehart H, Pang C, Pring B (2008) Market trends: Software as a service, worldwide, 2007–2012. Gartner report

Mertz S, Eschinger C, Eid T, Swinehart H, Pang C, Pring B (2009) Market trends: Software as a service, worldwide, 2008–2013, update. Gartner report

van Oosterhout M, Waarts E, van Hillegersberg J (2006) Change factors requiring agility and implications for it. Euro J Inf Syst 15(2):132–145

Panzar J, Willig R (1981) Economies of scope. Am Econ Rev 71(2):268–272

Papazoglou M (2007) Web services: Principles and technologies. Prentice Hall, Upper Saddle River

Porter M, Millar V (1985) How information gives you competitive advantage. Harv Bus Rev 63(4):149–160

Prahalad C, Ramaswamy V (2004) The future of competition: Co-creating unique values with customers. Harvard Business School Publishing, Boston

Rappa M (2001) Business Models on the Web. http://digitalenterprise.org/models/models.html, Accessed on 26 Apr 2011

Richardson L, Ruby S (2007) RESTful web services. O'Reilly, Sebastopol

Sääksjärvi M, Lassila A, Nordström H (2005) Evaluating the software as a service business model: From cpu time-sharing to online innovation sharing. In: Proceedings of the IADIS international conference e-Society, Qawra, pp 177–186

Sambamurthy V, Bharadwaj A, Grover V (2003) Shaping agility through digital options: Reconceptualizing the role of information technology in contemporary firms. MIS Q 27(2):237–263

Tapscott D, Ticoll D, Lowy A (2001) Digital capital: Harnessing the power of business webs. Harvard Business School Press, Boston

Part I
Web Service Business Models

Chapter 2
Services vs. Web Services

Without doubts, services have become the major driver of value creation in the last decades. This manifests in official statistics showing that services make up the largest part of the gross domestic product (GDP) in industrialized countries. In 2009, the share of the GDP within the European Union amounted to 71.9% and in the United States to 76.9%[1] increasing steadily over last years. This trend is further amplified by the "servicification" of traditional products in many industries. According to Vargo and Lusch (2004), the major shift towards a service-centered view is driven by changes in society and markets that lead to exchanges of services rather than goods. It is not only stagnant product demand in many domains, but also the customers' demand for customized and sophisticated goods which has pushed economic value downstream – away from manufacturing and toward the offering of services, both in preparing and customizing sales and in aftersales (Baumgartner and Wise 1999; Oliva and Kallenberg 2003). Driven by advancing Web service technologies, servicification in the software industry is a fundamental trend that tremendously changes the companies' strategies and business models: software vendors become service providers (Dubey and Wagle 2007). The growing importance of automated service provision over the Web is impressively documented by the rise of platforms like Salesforce.[2]

Chapter 2 discusses the general concept of a *service* from an information systems (IS) perspective. On the one hand, current economic definitions of services, mostly originating from the marketing discipline, are not clear-cut enough to transfer them to networks in which Web services are traded. Moreover, the definitions lack technical background. On the other hand, computer scientists have provided a bunch of definitions for (Web) services which are, however, too technical to tackle the business aspects. Hence, scholars did not yet close the gap between economists and computer scientists perspectives on services, which is, however, essential to grasp business aspects of Web services. Based on an extensive literature overview

[1] https://www.cia.gov/library/publications/the-world-factbook/fields/2012.html.

[2] http://www.salesforce.com.

C. Weinhardt et al., *Business Aspects of Web Services*,
DOI 10.1007/978-3-642-22447-8_2, © Springer-Verlag Berlin Heidelberg 2011

provided in Sect. 2.1, a definition and differentiation of the terms *service*, *electronic service*, and *Web service* is given in Sect. 2.2.

2.1 Service Definitions in Literature

Up to now, each research direction that dealt with service worked isolatedly within the bounds of a definition that suited the respective needs best. This approach led to an immense amount of diverse sector-specific definitions of "services." While computer scientists define a service based on the requirements emerging from practical implementations and, hence, focus on the technical properties of a service, they create a specification that economists cannot deal with. On the other hand, business economists put the emphasis on general properties of a service, its prerequisites, and the creation of value, thereby neglecting technical issues. In the following, an overview of existing business-related and technical service definitions is given.

2.1.1 Business Perspective

Considering explicit definitions, three perspectives on value creation are identified which most service definitions are based on: *potential-*, *process-*, and *outcome-*orientation Engelhardt et al. (1993). The former focuses on the allocation of factors of production and hence comprises the *preparation* for service supply. The second dimension considers services as the activation and integration of the allocated resources, that is services describe the *activity of resource usage* by both consumer and provider. The third dimension concentrates on the *result* of this process.

In line with Engelhardt et al. (1993), when defining a service, one needs to distinguish between its *preparation* as a first phase and its *delivery* as a second phase. The first phase includes general preparation activities that do not require the presence of the service customer herself or entities owned by her. For a hair cutting service, such activities would include, e.g., the education or hiring of staff. Analogously, for a Web-based service, programming efforts would be part of the preparation.

The first phase creates the readiness of a service provider, setting up the requirements for the immediate ability of a service provider to actually deliver a service. In case of a hair cutting service, such preparation would include, for instance, the education or hiring of staff, procurement of equipment, leasing of an accommodation, and so forth. For a Web-based service, the preparation phase would, inter alia, include programming efforts and the allocation of sufficient storage and computation capacity. In more detail, the preparation phase includes general preparation activities that do not require the presence of the service customer herself or entities owned by her.

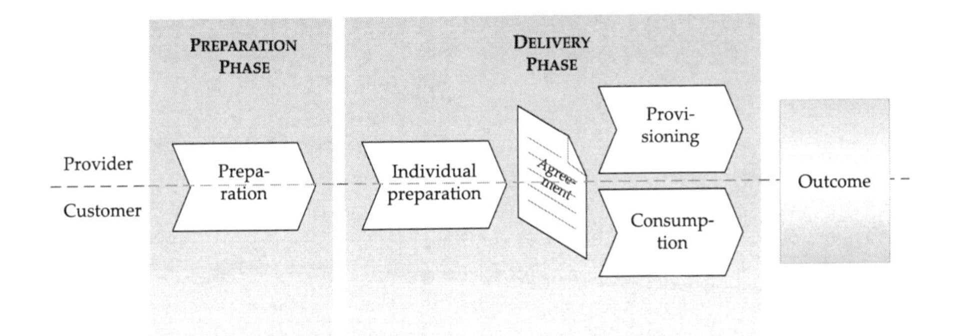

Fig. 2.1 The interrelation of service preparation, delivery, and its outcome

The delivery phase revolves around provisioning of the service by the provider and its simultaneous consumption[3] by the customer itself or entities owned by her. Within the phase of delivery, an *agreement* about the provision of a service is established between service provider and service consumer. Before starting the actual service delivery, *individual preparation* is carried out.[4] The preparation for service provision is universal, by means of representing the basis for service delivery in general. The outcome of a service delivery in contrast relates to a distinct service provision.

Additionally, further sub-phases need to be considered: Typically, an *agreement* between service provider and service consumer is made before provisioning and consumption is to be initiated, oftentimes including *individual preparation*.[5] While the preparation phase is universal, i.e., represents the basis for every concrete service delivery, the outcome relates to a distinct delivery process.

The phases introduced above are illustrated in Fig. 2.1.

This book takes the view that a potential-orientated definition of services is inappropriate since the allocation of factors is necessary in almost any value creation activity and therefore is not sufficiently distinctive. Outcome-orientation concentrates on the nature of the outcome being material or immaterial wherein, again, the essence of a service cannot be found. It is rather the delivery phase that defines the heart of a service by adequately and equally awarding importance

---

[3] According to the uno acto principle, delivery and consumption of a service are inseparable. Please refer to the inseparability characteristic of the IHIP criteria below.

[4] Services are subject to high degrees of customization (cf. the heterogeneity characteristics of the IHIP criteria below).Therefore, additional preparation in terms of customization is likely to antecede service provisioning and consumption. Hence, the more personalized a service is, the larger is the importance of the individual preparation included.

[5] Services are subject to high degrees of customization (cp. the heterogeneity characteristics of the IHIP criteria below). Therefore, additional preparation in terms of customization is likely to antecede service provisioning and consumption. Hence, the more personalized a service is, the larger is the importance of the individual preparation included.

to both the service provider and the service customer (Engelhardt et al. 1993). In order to prepare the service definition that underlies this chapter, a selection of the most frequently cited service definitions in academia which are focused on service delivery is presented in the following.

The *service dominant logic* is rather a philosophy than a definition. It considers services as "the application of specialized competencies (knowledge and skills) through deeds, processes, and performances for the benefit of another entity or the entity itself" (Vargo and Lusch 2004). This view of the service world is shared by the concept of service systems (Maglio and Spohrer 2008; Spohrer et al. 2008), which are dynamic value co-creation configurations of resources. These include people, organizations, shared information (i.e., language, laws, measures, methods), and technology.

This quite encyclopedic view of services comes at the price of concreteness: the vagueness of the service dominant logic does not allow for an explicit definition of what constitutes a service. The *continuum-based approach* takes a similar line by defining dimensions to characterize services. Its key message is, however, that there is no dichotomy between products and services: a clear-cut separation of services and products is therefore not possible (Shostack 1977; Chase 1981; Berry and Parasuraman 1991).

Besides taking the phases of service preparation and delivery into account, definitions of service in the literature focus on characteristics of service as compared to goods. These are **i**ntangibility, **h**eterogeneity, **i**nseparability of production and consumption and **p**erishability (IHIP criteria[6]) (Regan 1963; Rathmell 1966; Shostack 1977; Zeithaml et al. 1985; Edgett and Parkinson 1993).

Intangibility means that a service and its characteristics cannot be perceived before being bought (Kotler and Connor Jr 1977; Flipo 1988). Many scholars argue that intangibility is the central difference between products and services (Shostack 1977). *Heterogeneity* denotes the potentially high variability in service delivery (Zeithaml et al. 1985), by means of high degrees of customization, flexibility, and variability. Merging the perspective of service dominant logic to this issue, services that are performed with a particular input of specialized competencies, are likely to be subject to potentially varying performance of people (Rathmell 1966; Carman and Langeard 1980; Zeithaml et al. 1985). However, heterogeneity can also be interpreted as a source for providing high degrees of customization, flexibility, and variability. *Inseparability* reflects the so-called uno acto principle, that is the simultaneous consumption of services with delivery (Regan 1963; Wyckham et al. 1975) (cp. also phase 2 in Fig. 2.1). It also enables consumers to affect or shape the performance and quality of service (Grönroos 1978; Zeithaml 1981). *Perishability* circumscribes that services cannot be stored, inventoried, or transported (Rathmell 1966; Donnelly Jr 1976; Zeithaml et al. 1985).

The applicability of the IHIP criteria has been subject to discussion (cp. e.g., Lovelock and Gummesson (2004); Vargo and Lusch (2004b); Edvardsson et al.

[6]The IHIP literature is incidentally interconnected with the continuum-based approaches.

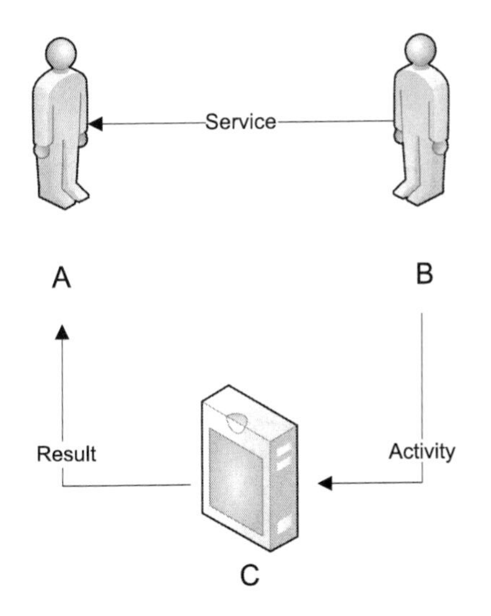

Fig. 2.2 Hill's service triangle

(2005)). Neither of the criteria allows for a precise distinction between services and goods, as there exist exceptions for each of them (e.g., car repair, programming, taking photos, etc., take a form of tangible representation). While recognizing that there are numerous services which take some form of tangible representation (e.g., car repair, programming, taking photos, etc.) services cannot be defined to be completely intangible. Analogously, there are products which are highly adapted to customers' requests (e.g., dell notebooks) and hence cannot be differentiated from goods by their heterogeneity.

A more detailed description of a service was proposed by Hill (1977). There, a service denotes an activity that is performed by an economic unit B for an economic unit A, where the result of this activity is the change in condition of an economic unit C that either is or belongs to economic unit A (cp. Fig. 2.2). Additionally, the prior agreement of economic unit A is assumed. Hill (1977) defines the "production of a service" as the activity itself which affects persons or goods. Hence, it is not the ability to perform a task (potential-oriented definition) but rather the performance itself that constitutes the "production of a service" and therefore, the creation of value. Transferred to the notation introduced in Fig. 2.1, Hill (1977) refers to the provisioning phase of a service.

The formalization of a service according to Hill (1977) specifies standard services like hair-cutting, car repairs, or post shipment very accurately. However, the definition does not allow for a distinction of services performed on electronic data or provided via electronic networks which is required for an interdisciplinary understanding of services.

Gadrey (2000) formulated several extensions to the definition provided by Hill (1977). Three necessary extensions are defined in order to include assistance and intervention, provision of technical capacities, and live performances in Hill's definition. However, the definitions of Hill (1977) and Gadrey (2000) do not include the concept of value co-creation by consumer and provider. Moreover, technical services do not only include the provision of technical resources, but also operations on electronic data or information. Thus, the definition has to be extended and stated more precisely in order to allow not only for goods and persons to be affected, but as well for electronic resources in general to be changed.

2.1.2 Computer Science Perspective

More recently, with the advent of service-oriented computing and Web services, perspectives from information and computer science increasingly entered the research field of service science. Technical services do not only affect humans and goods, but also other electronic resources such as addressable data sets. From a technological viewpoint, Web services are situated on an abstraction layer above different network protocols, operating systems, and programming languages. Thus, Web services provide possibilities to expose the functionality of an application system by means of Web technologies (Alonso et al. 2004).

Apart from this rather generic characterization and akin to the business-related definition, no agreement on a common definition on Web services has been made so far. According to W3C (2004), a Web service is a software system that is designed to support interoperable machine-to-machine interaction over a network. Part of the definition is also an interface, which is described in a machine-processable format – in particular, WSDL. Other systems must be able to interact with the Web service in a way that is prescribed by its description using SOAP-messages. The latter are transmitted via HTTP with an XML serialization in conjunction with other Web-related standards.

Berners-Lee et al. (1998) supplement the definition above by concretizing that a Web service is a software service, which is identified by a uniform resource identifier (URI), exposing a public interface based on Internet standards.

The definitions sketched above may serve computer scientists' needs, but do not include important economic aspects of a service such as involved parties besides electronic resources, changes of state, and value creation.

Based thereupon, the following section introduces a definition of service and a classification with respect to electronic and Web services, incorporating both the business perspective discussed in Sects. 2.1.1 and 2.1.2.

2.2 Classifying Services

The previous section introduced several definitions of *service*, each of which has a slightly different purpose. In order to define service and specify what are the characteristics of electronic and Web services, no delineation of services from goods is required explicitly. This sections takes a constructive approach. Both technical definitions aiming at the formal description of a service by means of exchanged messages, protocols, or interfaces and definitions that stress the economic aspect of services shall be amalgamated in order to introduce a general definition of services that holds for interdisciplinary service research. Build thereupon, special types of services are classified that are considered frequently in the emerging fields of SOA and service composition (Papazoglou and Dubray 2004).

Based on Hill (1977) and Gadrey (2000), a service shall be defined as follows:

Definition 2.1 (Service). A service is a set of activities performed and intended to bring about a change of state of either an entity that is owned or used by a consumer or the consumer itself. The set of activities are performed by a provider or jointly by provider and consumer. The resulting change is based on a prior agreement between consumer and provider which aims at a co-creation of value.

The above definition stipulates the intention to change the state of an entity, e.g., cut hair or a higher degree of information of students. Furthermore, it requires a prior agreement on the provision of the service in order to exclude unrequested acting. By co-creation of value, contribution of both parties, consumer and provider, is understood. The customer's contribution can be manifold, reaching from provisioning of the entity that is changed to intensely supporting the execution of the set of activities. The provider's contribution is the performance of the activities that constitute the provisioning of a service. Definition 2.1 shall be able to cover any kind of service, ranging from hair cutting, to teaching or consulting, to entirely Web-based services such as, for instance, Amazon Web services.

As information and communication technology rises and the Web grows rapidly, the environment of service provision changed completely. In this context, a special kind of *service* emerged that is defined as *Electronic Service*.

Definition 2.2 (Electronic Service). An electronic service (e-Service) is a service of which the input and output are provided via an electronic network.

In more detail, a service turns into an electronic service as soon as both the input provided by the customer and the output of the service delivery take place via electronic channels. The latter can, for instance, be a telephone network or the Internet. A service that is entirely based on telephone communication is classified to be an electronic service just like services that are initiated and finalized via email communication or Web protocols. Note that the actual activity performed by the provider is not required to be electronic, however, any kind of interaction, as for instance follow-up inquiries that take place within service delivery, must be made via an electronic network.

The undergoing change in the software industry brings about on-demand services that are entirely hosted and maintained by the service provider and can be accessed via Web browsers. Due to their well-defined interface, they are interoperable with other services of that kind, being composable to meet individual customer requests. This specific kind of e-service is denoted a Web service, which requires a renewed specialization of Definition 2.2.

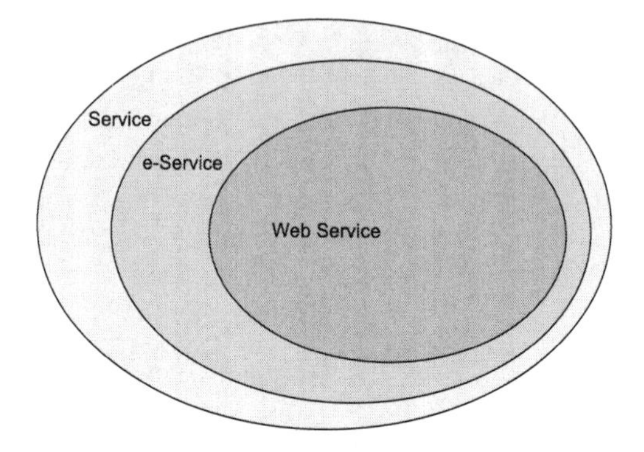

Fig. 2.3 Service categories and their interrelation

Definition 2.3 (Web Service). A Web service is an e-service identified by a URI that exposes a public, well-defined interface. The input and outcome or output are provided via a Web protocol.

The differentiation of a Web service and the other aforementioned service types is twofold. First, the requirement of having a URI that exposes a publicly available, well-defined interface must be included (as explained in Sect. 2.1.2). A well-defined interface is not only essential from a computer science perspective, but also lays the groundwork for joint value creation in service networks. Second, Definition 2.2 is specialized in terms of the nature of how the input and outcome must be provided. Typical examples for Web services are salesforce.com's Sales Cloud[7] or Amazon Web services,[8] e.g., EC2 or S3. Yet, akin to Definition 2.2, service provisioning itself does not need to take place via communication by means of a Web protocol. A prime example for such a Web service is Amazon's Mechanical Turk,[9] an on-demand scalable workforce in which inquiries and the output are standardized as in any "traditional" Web service, however, the contents that are requested are processed by humans. Figure 2.3 summarizes the definitions provided in this section.

2.3 Summary

In this chapter, an overview of existing service definitions was presented in order to illustrate the different perspectives taken on services. Analyzing the different definitions, it becomes obvious that no definition exists yet that can be applied in the field of technical service composition as well as in economic considerations.

[7]http://www.salesforce.com/crm/sales-force-automation/.

[8]http://aws.amazon.com/.

[9]https://www.mturk.com/mturk/.

Table 2.1 Criteria of service categories

Attribute	Service	e-service	Web service
Input/output transmission	Electronically/personally	Electronically	Electronically
Interface	(Not) well defined	(Not) well defined	Well defined
Communication protocol	Any	Any	Web protocol

Therefore, Sect. 2.2 provided a general definition of services. Distinctive characteristics for specialized types of electronic services and Web services were identified resulting in a description for each service type.

According to the characteristics, which are illustrated in Table 2.1, an electronic service is a service with an electronic transmission of input and output. Analogously, a Web service is an e-service that exposes a well-defined interface and where communication is regulated by means of a Web protocol.

Web services build the basis for "service value networks," which are introduced in the following chapter.

References

Alonso G, Casati F, Kuno H, Machiraju V (2004) Web services: Concepts, architectures and applications. Springer, Berlin

Baumgartner P, Wise R (1999) Go downstream – the new profit imperative in manufacturing. Harv Bus Rev 77(5):133–141

Berners-Lee T, Fielding R, Irvine U, Masinter L (1998) Rfc2396: Uniform resource identifiers (uri): Generic syntax. Tech. rep., The Internet Society

Berry L, Parasuraman A (1991) Marketing services: Competing through quality. Free Press, New York

Carman J, Langeard E (1980) Growth strategies for service firms. Strat Manag J 1(1):7–22

Chase R (1981) The customer contact approach to services: Theoretical bases and practical extensions. Oper Res 29(4):698–706

Donnelly J Jr (1976) Marketing intermediaries in channels of distribution for services. J Market 40(1):55–57

Dubey A, Wagle D (2007) Delivering software as a service. Online, URL http://www.mckinsey.de/downloads/publikation/mck_on_bt/2007/mobt_12_Delivering_Software_as_a_Service.pdf, the McKinsey Quarterly, Web exclusive

Edgett S, Parkinson S (1993) Marketing for service industries – a review. The Serv Indust J 13(3):19–39

Edvardsson B, Gustafsson A, Roos I (2005) Service portraits in service research: a critical review. Int J Serv Indust Manag 16(1):107–121

Engelhardt W, Kleinaltenkamp M, Reckenfelderbäumer M (1993) Leistungsbündel als absatzobjekte. Zeitschrift für betriebswirtschaftliche Forschung 45(5):395–426

Flipo J (1988) On the intangibility of services. Serv Indust J 8(3):286–293

Gadrey J (2000) The characterization of goods and services: An alternative approach. Rev Income Wealth 46(3):369–387

Grönroos C (1978) A service-orientated approach to marketing of services. Eur J Market 12(8):588–601

Hill T (1977) On goods and services. Review of Income and Wealth 23(4):315–338

Kotler P, Connor Jr R (1977) Marketing professional services. J Market 41(1):71–76

Lovelock C, Gummesson E (2004) Whither services marketing? in search of a new paradigm and fresh perspectives. J Serv Res 7(1):20–41

Maglio P, Spohrer J (2008) Fundamentals of service science. J Acad Market Sci 36(1):18–20

Oliva R, Kallenberg R (2003) Managing the transition from products to services. Int J Serv Ind Manag 14(2):160–172

Papazoglou M, Dubray J (2004) A survey of web service technologies. Tech. rep., University of Trento, Department of Information and Communication Technology

Rathmell J (1966) What is meant by services? J Market 30(4):32–36. URL http://www.jstor.org/pss/1249496

Regan W (1963) The service revolution. J Market 27(3):57–62

Shostack G (1977) Breaking free from product marketing. J Market 41(2):73–80

Spohrer J, Anderson L, Pass N, Ager T, Gruhl D (2008) Service science. J Grid Comput 6(3):313–324

Vargo S, Lusch R (2004a) Evolving to a new dominant logic for marketing. J Market 68(1):1–17

Vargo S, Lusch R (2004b) The four service marketing myths: Remnants of a goods-based, manufacturing model. J Serv Res 6(4):324–335

W3C (2004) Web services glossary. http://www.w3.org/TR/2004/NOTE-ws-gloss-20040211/

Wyckham R, Fitzroy P, Mandry G (1975) Marketing of services. Eur J Market 9(1):59–67

Zeithaml V (1981) How consumer evaluation processes differ between goods and services. In: Donnelly J, George W (eds) Marketing of services. American Marketing Association, Chicago, pp 186–190

Zeithaml V, Parasuraman A, Berry L (1985) Problems and strategies in services marketing. J Market 49(2):33–46

Chapter 3
Service Value Networks

Following the service definition in Chap. 2, the organizational form of *service value networks (SVNs)* will be introduced as a novel network type and specialization of business networks. Section 3.2 contributes a fundamental definition of the SVN concept and its differentiation from related network types. As it is shown, however, SVNs exhibit special characteristics compared to the known definitions that are not yet sufficiently discussed and formulated. As both academics and practitioners still lack approaches to formalize and economically analyze SVNs, this research gap is filled by Sect. 3.3. It introduces a formalization of SVNs that will serve as the notational basis for the mechanism design approaches introduced in Chap. 7 of this book. Several examples for SVNs presented in Sect. 3.4 round off this chapter.

3.1 Introduction

Business models of companies that operate in the software industry as well as the fashion they create value are radically changed by tremendous technical and economic developments. While the software industry used to earn money with large and monolithic software applications implying license-based pricing models, the on-demand delivery of flexible service components, which are operated in service-oriented architectures (SOA), begin to outstrip the traditional business model. Modular components based on Web service technology can easily be extended by additional services. License-based full versions based on upfront payments are being replaced by software as a service (SaaS). This is not only a technical evolution, but also a conceptual one: customers are given the opportunity to purchase services on-demand and, thereby, to move away from the over-dependance on one single vendor. Service modules can rather be composed from the offerings of different providers. The services' modularity allows for a tight focus on required features.

Such modularity is one of the most promising answers to the question of how to face rising demands for sophisticated, customized services (Baldwin and Clark 2000). Once serving the whole value chain by what has become famous as vertical

C. Weinhardt et al., *Business Aspects of Web Services*,
DOI 10.1007/978-3-642-22447-8_3, © Springer-Verlag Berlin Heidelberg 2011

integration, service providers now tend to engage in networked value creation in ecosystem-like environments. Through horizontal specialization, service providers leverage the knowledge and capital assets of partners and lower the risk of operating in a changing and uncertain environment. This chapter shall precisely define SVNs, thereby laying the ground for a rising research area on a new business trend.

Management literature, social and computer science have developed tons of definitions for all kinds of networks. However, SVNs exhibit special characteristics compared to the known definitions.

3.2 Definition and Related Concepts

Academic literature has failed to provide an explicit and unambiguous definition of SVNs to date (cp., e.g., Hamilton (2004); Chesbrough and Spohrer (2006); Hamilton (2007); Basole and Rouse (2008); Bitsaki et al. (2008)). This section shall map out the conceptual boundaries explicitly formulate a definition of SVNs based on a comparison to related concepts from organizational theory. The analysis will show that many of their characteristics overlap. Still, distinctive features can be found.

Ever since the seminal work of Williamson (1973), the three basic types of business governance structures market, hierarchy, and the hybrid networks "in between" have established. For the one extreme position, pure markets, information is said to be publicly and instantaneously observable through the market price. In such an ideal environment, the price is a perfect incentive mechanism which aligns the market participants' profits with economic efficiency. Sunk costs through high factor specificity, uncertainty in demand and supply and, thus, frequency of trade, and high customization requirements turn the assumption of an effective price mechanism unrealistic for non-standardized goods and services. In addition, the knowledge-based view of the firm points out that a firm's knowledge and core competencies cannot be appropriately protected by a purely market-based organization. On the contrary, purely hierarchical organizations create value strictly within the firm's boundaries. Such formal control, by definition, rules out opportunistic behavior of business partners. This comes, however, at the price of inefficiency and inflexibility – an almost certain death-blow in today's increasingly competitive and quickly changing business environments. This holds particularly true in ICT-related domains with ever shorter innovation cycles.

The most common hybrid governance form "network" combines the advantages of market governance, most strikingly adaptability and efficiency, with the oftentimes indispensable requirements control and protection of knowledge and core competencies yielded by hierarchies. Consequently, business networks have been proposed as the superior governance form for today's highly dynamic and complex business world (Miles and Snow 1986; Thorelli 1986; Baker 1992).

In summary, the most important advantages of networked organizations compared to more traditional governance forms are as follows:

- Insurance against uncertainty in supply and demand (Jones et al. 1997).
- Balance adaptability, even with respect to complex tasks, while control is maintained (Hamel et al. 1989; van Heck and Vervest 2007).
- Protection of business knowledge through modularization (Baldwin and Clark 2000; Hoogeweegen and Vervest 2005; Hoetker 2006).
- Market-based coordination to ensure efficiency (Miles and Snow 1986).

A bulk of managerial and academic literature deals with business networks (BNs) (cp., e.g., Holm et al. (1996); Hakansson and Ford (2002); Ritter et al. (2004)) and variations thereof. Primarily, these are strategic alliances (Devlin and Bleackley 1988; Yoshino and Rangan 1995), virtual organizations (VOs) (Goldman et al. 1995; Holland and Lockett 1998), and smart business networks (SBNs) (Vervest et al. 2004, 2005; van Heck and Vervest 2007, 2009; Busquets et al. 2009). Clearly, SVNs should also be on this list. But what distinguishes SVNs from the other listed organizational types?

Business networks denominate the most universal form of a cooperation among firms or legal entities that is economically motivated (Holm et al. 1996). As quintessentially for any cooperation, participation in a BN is based on the sentience of mutual benefit and aimed at the co-creation of business value. Literature proposes that business networks tend to be either temporary, project-driven, or goal-oriented. Network "partners" may both be competitors and complementers (that is, providing heterogeneous goods or services) (Bengtsson and Kock 2000).

Strategic alliances and *virtual organizations* are derivatives of business networks. Strategic alliances mainly denote cooperations among firms that are competitors outside of this kind of business network. The intention of a strategic alliance is to share risk or achieve economies of scale (Hamel et al. 1989; Mowery et al. 1996). Virtual organizations, on the contrary, stress the cooperation of firms that bring in complementary competencies and resources for mutual benefit (Davidow and Malone 1993). VOs enable the partners to achieve a whatsoever goal one alone would not be able to master (Strader et al. 1998; Wolters and Hoogeweegen 1999).

Smart business networks emphasize the *smart* use of ICT to facilitate network interaction within BNs. Hereby, smartness is a relative term denoting comparative advantage and effectiveness through the use of ICT (van Heck and Vervest 2009). At the same time, ICT is considered an enabler for network agility, that is, the SBN's ability to "rapidly pick, plug, and play" business processes (van Heck and Vervest 2007). The concept of SBNs closely relates to the shift from "mass customization" to "mass individualization" since the postulated agility allows for a flexible adaptation to quickly evolving customer requirements (Busquets et al. 2009). On the other hand, as soon as an individual request has been completed, the SBN can be easily "disintegrated." These quick connecting and disconnecting capabilities are facilitated through the modular provision of business capabilities by the partners in the smart business network. This modularity of network members also is also suited for the protection of the partner's core competencies (Hoogeweegen and Vervest

2005): Trust problems, which are commonly encountered in purely market-based organizational forms can be diminished, at the same time the open pool of partners in SBNs allows for a the necessary flexibility of value creation (Tiwana 2008; van Heck and Vervest 2009). Value creation requires a "network coordinator" (Hinterhuber 2002), which can either be determined (e.g., through the role of a network operator) or coincidence.[1]

This review shows that the same prototypical characteristics cited by the service science community to describe SVNs are also found in the more general organizational concepts listed above. Spohrer et al. (2008) notes the interaction of different entities for mutual benefits; Hamilton (2007) addresses the existence of and complementing participants in SVNs. Yet, both characteristics are also a parts of business networks in general. Basole and Rouse (2008) stress the role of ICT as an enabler for SVNs, moreover, the customers are seen as the "triggers of all activities in the network." However, looking at various definitions of SBNs, these elements are to be found there, too.

Another differentiation criterion of SVNs could be the explicit link to the service domain. As this book does not seek to differentiate services from goods (cp. Sect. 2.1) and service researchers themselves postulate that products can be seen as "vehicles for service delivery" (Vargo and Lusch 2004; Araujo and Spring 2006), such a contour does not seem to be reasonable either.

Much more precisely and concrete, Definition 3.1 specifies SVNs are a subset of SBNs that perform the coordination and orchestration of services automatically via a universally accessible network orchestration platform. This platform itself does not have to be technically centralized (cp. also Fig. 3.1). Table 3.1 summarizes the definition variations and the resulting changes in governance structures.

Definition 3.1 (Service Value Network). SVNs are SBNs that provide business value by performing automated on-demand composition of complex services from a steady but open pool of complementary as well as substitutive standardized service modules through a universally accessible network orchestration platform.

Definition 3.1 is distinctive since network composition and network orchestration is not automated in SBNs by definition. In the following, each component of the definition is discussed in more detail.

Complex services. A complex service is composed of various component services. In more detail, complex services typically involve the assembly and invocation of several component services offered by diverse enterprises in order to complete a multi-step business functionality (Papazoglou 2007). *Component services* are either other complex services or "atomic" functionalities provided via a Web service.

Standardized service modules. In order to be plug-and-playable, service modules need to provide standardized interfaces. Additionally, the automated service

[1]An archetypal example of a network orchestrator within an SBN is Li and Fung (http://www.lifung.com/eng/global/home.php), which coordinates a network of more than 8,000 network partners.

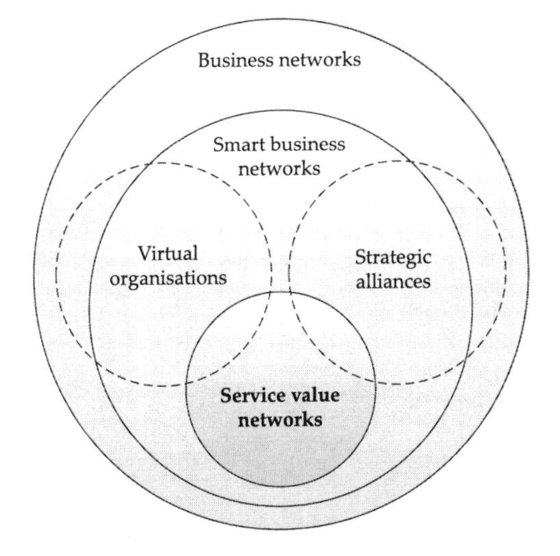

Fig. 3.1 Different concepts of networked organizations and their interrelations

Table 3.1 Differentiation between network types

Network type	Distinguishing characteristics
Service value network	Interoperable service modules
	Automated service orchestration
Smart business network	Non-automated service orchestration
Business network	Non-compatible service modules

orchestration within SVNs requires that standardization encompasses the specification of acceptable parameter values in machine-readable form.

Steady but open pool of complementary and substitutive services. Services must register or be registered (either automatically[2] or manually) with the SVN in order to be allowed for composition. The set of registered services forms the steady pool from which a complex service is composed. Yet, an SVN may not be a closed network: registration is open for any service that meets certain minimum requirements, for instance, as to appropriate interface specifications. *Steady* denotes a list of services (including their interface descriptions) maintained by the SVN – which is available also in case of no current service composition request in the network. *Openness* means that no service can be excluded from the network, as long as it meets the publicly known minimum requirements defined by the platform provider.

Automated on-demand service composition. On-demand service composition refers to the SVN's ability to orchestrate a complex service in an ad-hoc fashion and

[2]For instance, via an active browsing of the service landscape for eligible services performed by the SVN.

upon customer request. Importantly, at the time of the request, the SVN will *automatically* seek an optimal path through the network of registered services. Generally, a path through the network maximizes the economic surplus of one or more network participants. Thereby, optimality is evaluated in terms of a goal function, for example the overall wealth of all involved parties (*market efficiency*). An automation of the complex service orchestration process does not only require a goal function, but also a mechanism that is able to maximize it – as, for example, the mechanisms presented in this book (cp. Sect. 7).

Universally accessible network orchestration platform. Eventually, the above SVN components are brought to life via the network orchestration platform. The platform spans the business logic and technical infrastructure which is necessary to perform market-based on-demand service composition. The platform also maintains the above-mentioned list of orchestratable (and registered) services. As such, the platform is the interface for and between service customers and providers. Consequently, it is a crucial factor to the platform to grant universal access and permanent availability. If this merit cannot be assured, registered services may be excluded from the platform (thereby violating its openness requirement) and business opportunities may be lost.

3.3 Formalization of Service Value Networks

The automated composition of complex services forms the nucleus of SVNs. In the following, a mathematical formalization of SVNs will be provided on whose basis the composition of complex services can be performed automatically.

3.3.1 Formal Model

For the sake of clarity, the focus is on a particular instantiation of the overall SVN, including only the services that fulfill the functionality demanded by a specific customer request, rather than tackling the entire SVN. The following model of an instantiated SVN is described by means of a simplified statechart model (Harel and Naamad 1996) and is aligned with the representation presented by Yang (2003). Statecharts have proven to be the preferred choice for specifying process models because they expose well-defined semantics and provide flow constructs from prominent process modeling languages (e.g., WS-BPEL), which allows for simple serialization in standardized formalisms (cp., Fig. 3.2).

A service value network is represented by a directed, k-partite, and acyclic graph. Each partition represents a different functionality requested by the service customer. The set of nodes $V = \{v_1, \ldots, v_n\}$ represents the set of service offers that can fulfill the requested functionality. Two auxiliary notes, source (v_s) and sink (v_f), act as a makeshift to formalize complex services as an end-to-end connection. These nodes should therefore not be interpreted as services in the network.

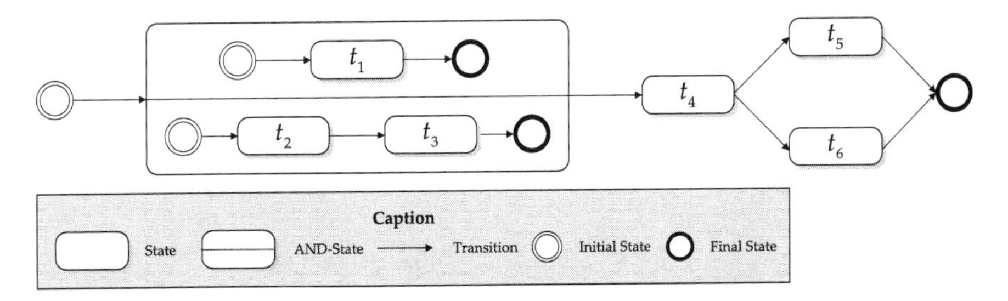

Fig. 3.2 Statechart formalization (Harel and Naamad 1996; Yang 2003)

Services are offered by a set of m *service providers* $N = \{n_1, \ldots, n_m\}$ with n being an arbitrary service provider. The *ownership information* $\sigma : N \rightarrow \mathscr{P}(V)$ that reveals which service provider owns which services within the network is public knowledge.[3]

According to the different service functionalities that are combined in order to constitute the demanded complex service, services are clustered into k partitions – the so-called candidate pools. Substitutive services are mapped to one and the same candidate pool. The vector of all candidate pools is denoted by $Y = (Y^1, \ldots, Y^{\tilde{k}}, \ldots, Y^k)$, $1 \leq k \leq |V|$ with $Y_{\tilde{k}}$ being an arbitrary candidate pool. Candidate pools specify the sequence $[1, \ldots, k]$ of functional steps within a complex service. Exactly one service out of each candidate pool is required to deliver an instance of the complex service requested by the customer. Let $v_j \in V$ be an arbitrary service in the network. Let $v_j^{\tilde{k}}$ denote that service v_j belongs to candidate pool $Y^{\tilde{k}}$. Source and sink are not considered a separate candidate pool. However, for notational simplicity the virtual candidate pools $Y^0 = \{v_s\}$ and $Y^{k+1} = \{v_f\}$ are introduced, which contain the source and sink, respectively.

An edge $e_{i^{\tilde{k}} j^{\tilde{k}+1}}$ denotes a composition relationship between the nodes $v_i^{\tilde{k}}$ and $v_j^{\tilde{k}+1}$. In other words, an edge between two nodes symbolizes the interoperability of the services offered as well as the service providers' willingness to cooperate. In order to represent the sequence of functional steps, the set of edges is restricted such that edges link services from subsequent candidate pools, i.e., $E := \{e_{i^{\tilde{k}} j^{\tilde{k}+1}}\}, \forall \tilde{k} = 0, \ldots, k$ is the set of all possible links related to the set of service providers.

Each service v_j exhibits a *service configuration* $\mathscr{A}_j$ that is characterized by a vector $\mathscr{A}_j = \{a_j^1, \ldots, a_j^m\}$ where $a_j^{\tilde{m}}$ is an attribute value of the $\tilde{m}$th attribute type of service v_j, thereby unambiguously defining all relevant service characteristics. Furthermore, let p_{ij} denote the price for service v_j when being allocated as the successor of service v_i. Let c_{ij} denote the *internal variable costs* that the service provider that owns service v_j has to bear for that service being interoperable

[3]The reverse ownership information $\sigma^{-1} : V \rightarrow N$ maps service offers to single service providers that own that particular service.

with service v_i and for the execution of service v_j as a successor of service v_i. The representation of a detailed cost structure of service providers is intentionally omitted which serves a better understanding and does not restrict the generalization of the model. It is assumed that the representation of internal variable costs reflects the service providers' valuations for their service offers being executed in different composition-related contexts.

Definition 3.2 (Service Configuration). A service configuration $\mathscr{A}_j$ of a service $v_j \in V$ selects a value $a_j^{\tilde{m}}$ for each attribute type $\tilde{m}$ of a service and thereby unambiguously defines all relevant service characteristics. The choice of configuration might affect the functional and non-functional aspects of a service and is a major determinant of the price.

Thus, a concrete SVN can be formally described as the graph

$$G := (\{V \cup \{v_s, v_f\}\}, E) \text{ with}$$

$$E := \{e_{i^{\tilde{k}} j^{\tilde{k}+1}}\}, \forall \tilde{k} = 0, \ldots, k$$

Only feasible realizations of complex services, i.e., complete paths from source to sink, create value. It is important to note that a complex service incorporates exactly one service from each candidate pool. Thus a feasible instantiation of a complex service is a path $F_{\tilde{r}} = (e_{s^0 i^1}, \ldots, e_{i^{\tilde{k}} j^{\tilde{k}+1}}, e_{j^{\tilde{k}+1} l^{\tilde{k}+2}}, \ldots, e_{i^k f^{k+1}})$, $\forall \tilde{k} = 1, \ldots, k$.

Figure 3.3 illustrates an exemplary formalization of a service value network with two candidate pools. Every feasible path from source to sink represents a possible realization (instance) of a complex service. There are three paths $F = \{F_1, F_2, F_3\}$ with $F_1 = (e_{s1}, e_{13}, e_{3f})$, $F_2 = (e_{s1}, e_{14}, e_{4f})$, and $F_3 = (e_{s2}, e_{24}, e_{4f})$.

3.4 Assessing the Value of SVNs in Practice

As the organizational concept of SVNs has been defined and formally introduced in the preceding sections, it is time to consider different implementations of SVNs. Building on the basic structure of today's Web service market, the most prominent forerunner of SVNs – salesforce.com's AppExchange (cp., Sect. 3.4.2) – and TEXO[4] as a current and closely related research project (cp. Sect. 3.4.3) are discussed. Finally, fictional examples for complex services as they could be offered in a representative SVN, for example, in TEXO, will be presented.

---

[4]http://theseus-programm.de/en-us/theseus-application-scenarios/texo/default.aspx.

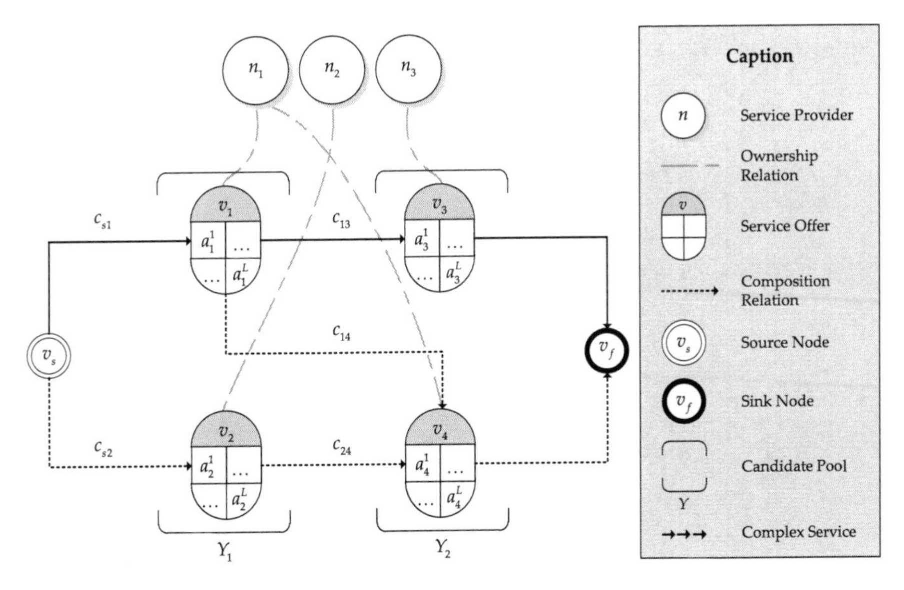

Fig. 3.3 Exemplary formalization of a service value network

3.4.1 The Web Service Market

The current Web service market points the way towards SVNs. The current market can be divided into four quadrants as shown in Fig. 3.4.

Figure 3.4 classifies the manifestations of the Web service market by the *complexity of the underlying business process* and the *degree of cross-organizational interaction* when developing and offering services. The variety of Web services that support single steps of a process offered by individual service providers is already vast: Prime examples are services provided by Google such as Google Docs[5] as a Web-based offering to create and share work online or Google Maps,[6] a map service that can be easily embedded in Web sites or service mashups. Other examples that enjoy high popularity are the Amazon Simple Storage Service (S3) as a simple Web service that can be utilized to store and retrieve data and the Amazon Elastic Compute Cloud (EC2) that provides resizable compute capacity in the cloud. These Web services are already extensively used in service mashups, as for instance reflected in ProgrammableWeb's composition matrix: Google Maps, for instance, is currently part of approximately 2,200 mashups listed at ProgrammableWeb while Amazon's S3 and EC2 are included in more than 100 mashups.[7]

[5]http://docs.google.com.

[6]http://maps.google.com.

[7]http://www.programmableweb.com/apis, Status: 03/21/2011.

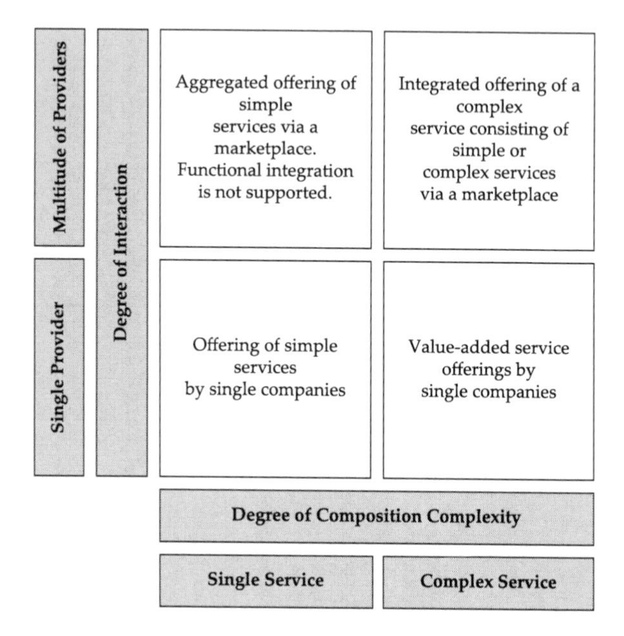

Fig. 3.4 Typology of Web services

On the other hand, applications supporting multi-step business processes are increasingly offered as Web services, too. For instance, companies like salesforce.com or Netsuite Inc.[8] successfully entered the business software market with their entirely Web based on-demand customer relationship management (CRM) suites. Components offered within these suites can be dynamically composed to customized processes. Additionally, traditional players in the software market started to enhance their business models towards Web-based offerings. SAP's enterprise resource planning application BusinessByDesign[9] and CRM on demand by Oracle[10] are only two examples for the postulated shift from traditional software products to services (Dubey and Wagle 2007). Yet, these services are provided by single vendors.

The field of complex services that are *composed of elements provided by different vendors* is where SVNs are to be classified. Web service marketplaces such as StrikeIron and Xignite can be interpreted as SVN forerunners: they do offer a platform where service providers can market their specialized Web services to customers, yet automated mediation is not available. For instance, neither functional integration nor automatized matchmaking are provided. Added value through service composition can by created, however, both choice of an optimal combination and the actual composition is left to the customer. Salesforce.com's AppExchange

[8]http://www.netsuite.com.

[9]http://www.sap.com/bydesign.

[10]http://www.oracle.com/crmondemand.

goes one step further into the SVN direction, providing a market for diverse vendors to offer their complementary services to Sales Cloud 2. This network exhibits some, but not every feature of a service value network (cp. also Sect. 3.4.2). The TEXO platform as a part of the TEXO research project[11] is a current example for providing both technical and economic support for SVNs as defined in this book.

3.4.2 Real-World SVN Forerunner: AppExchange

Already being the worldwide leader in on-demand CRM services, salesforce.com launched their marketplace for third party on-demand applications – AppExchange – in 2005 in order to innovate and extend their business model. The core idea of AppExchange is to offer a platform for complementary services grouped around Salesforce.com's core offering Sales Cloud 2 (formerly known as Salesforce CRM) in order to increase its value and range of covered functionalities. The bulk of services to be found at AppExchange are third-party applications offered by both freelance software developers and software companies. By March 2011, AppExchange included more than 1,100 Web services to complement Sales Cloud 2.[12]

The offered services are fully integrated into Sales Cloud 2 which slashes adjustment efforts on customer side. Seamless and automated integration is enabled by the fact that all services offered at AppExchange (1) are deliberately built for Sales Cloud 2 and (2) are based on and restricted to a proprietary but open development platform – Force.com.[13] Force.com as "platform-as-a-service" (PaaS) provides a means that allow developers to engineer applications that are pre-integrated in Sales Cloud 2 and can thus be marketed via AppExchange. Applications that are built via Force.com use the same proprietary programming language (Apex) and the same syntax to create the service's interface (Visualforce). With Force.com, an on-demand programming environment to facilitate third-party-built applications complements the portfolio offered. However, even if the customer assembles an integrated, complex service via AppExchange, single service level agreements (SLAs), billing, and a proprietary pricing scheme applies for each of the purchased component services. Pricing is thereby entirely left to the service providers: pricing structures range from a fee per user and month bound to an annual subscription to a fixed fee per year plus additional charges that are linked to present Sales Cloud 2 users, just to name two examples. Dynamic pricing as well as a pricing at a complex service level is not available. Moreover, as one of the most important features of SVNs, automated on-demand service composition is not featured. As it was shown in the previous section, AppExchange can therefore be classified an SBN. However,

[11]TEXO is one component of the umbrella research program THESEUS initiated by the German Federal Ministry of Economy and Technology. THESEUS aims at developing a novel Internet-based infrastructure to improve the usage of knowledge available on the Internet.

[12]The figures are taken from http://sites.force.com/appexchange/home/, Status: 03/21/2011.

[13]http://www.salesforce.com/platform/.

the applied enabling technology theoretically allows for an extension towards an SVN.

In total, Salesforce.com's AppExchange can certainly be rated the most prominent SVN forerunner with approximately 80,000 customers that are given access to a multitude of third-party services that provide complementary functionality to salesforce.com's core offering.[14] Applications are offered by approximately 450 diverse vendors, which shows that the postulated trend towards highly specialized providers that rely on their core competency (Baldwin and Clark 2000; Downes and Mui 2000; Tapscott et al. 2001) has already become reality in the Web services industry.

3.4.3 Research Activities on SVNs: TEXO

The TEXO project envisages the creation of a platform and marketplace that enables the innovation, engineering, offering, and consumption of Web services via the Internet supported by its underlying IT infrastructure. The focus of the services traded via TEXO is put upon value-added complex services that involve diverse modular service components offered by different providers. Thus, TEXO is an electronic service broker that allows for the offering and merchandizing of (complex) Web services, bringing together service supply and demand. The approach is based on a service oriented architecture with its capability of exposing and connecting single Web services from different sources (Barros and Dumas 2006).

Figure 3.5 gives a high level architecture overview on the TEXO platform components (Kabzeva et al. 2009). The architecture is aligned to the fundamental modeling concepts (FMC).[15]

The TEXO Service Management Platform (SMP) as the heart of TEXO comprises TEXO Tools, the TEXO Portal, TEXO Management Services, and data storage components. The TEXO Tools include two main components: the Innovation Cockpit and the Integrated Service Engineering (ISE) Workbench. The Innovation Cockpit offers tool-based support to service innovators such as idea mining and idea evaluation tools (Stathel et al. 2008). The ISE Workbench is a service unified development environment which gives providers support in modeling, design, and description (Scheithauer et al. 2009). Akin to Force.com, the ISE workbench lays down the minimal requirements that must be met in order to offer (compatible) services within the platform.

TEXO Management Services offer core platform functionalities including service registration, service discovery, complex service auctioning, service level manager, billing, and monitoring. These services, either offered by the platform itself or by third-party providers, are deployed to the TEXO Management Runtime to be accessible by other components, and are offered to service customers and

[14] https://www.salesforce.com/company/investor/financials/, Status: 03/21/2011.

[15] http://www.fmc-modeling.org/home.

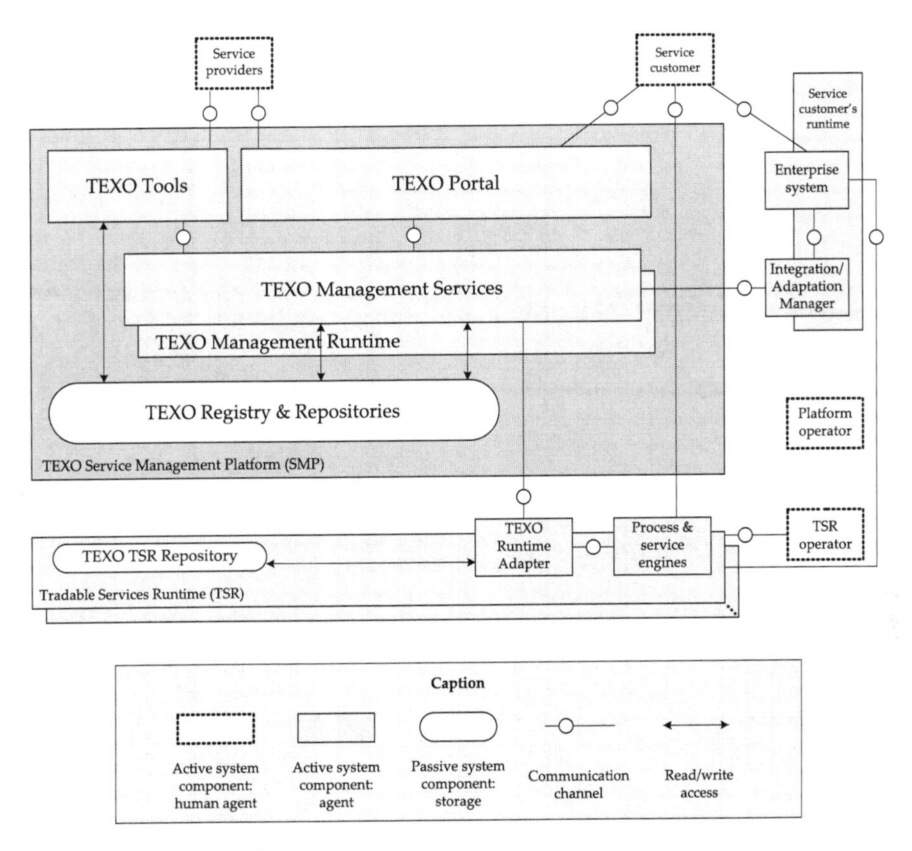

Fig. 3.5 High level TEXO architecture

providers via a Web-based frontend – the TEXO portal. Most importantly, the TEXO Management Services allow for an automated on-demand service composition that is a required feature in SVNs. Further, a comprehensive service level management, monitoring, billing, and pricing on complex service level is facilitated.

The TEXO registry allows for a registration of tradable services on the SMP while the repositories provide and store all required information on tradable services such as their description or monitoring data.

The required input to administrate monitoring and billing of services is provided by the Tradable Services Runtime (TSR). Any service hoster can install a TSR and offer a TEXO-related hosting service. As services are usually not hosted by TEXO itself, the TEXO runtime adapter provides for an exchange of required data between the TSR and the TEXO SMP. If a service provider does not host its service itself, it can choose a TSR to deploy its service. The TEXO TSR repository of each TSR stores the information of services that are deployed on the respective TSR. Locally installed process and service engines are required to deploy different service aspects, such as service process, or user interface.

Since an automated integration of services in the customer's enterprise system is another objective of the TEXO project, TEXO offers the Integration/Adoption Manager. It is responsible for managing all integration aspects of a tradeable service into the customer's service-based application, or enterprise system, respectively (Kabzeva et al. 2009).

The above-described architecture provides one possibility of a technical basis for SVNs which goes beyond the composition of services that are closely related to a core application as it is the case for AppExchange. In line with a service value network's characteristics, the TEXO platform is able to orchestrate a complex service on customer request with the help of the TEXO Management Services including automatized search for an optimal combination of services to be composed according to the demanded functional and non-functional customer requirements. The co-opetition mechanism elaborated in this thesis is a possible instantiation of a market mechanism that can be applied within the TEXO SMP.

3.4.4 *Fictional Examples of Complex Service Compositions*

Let us now turn to fictional examples based on currently available service offerings. An ideal SVN offers a Web interface that is capable of automatically locating and combining the individual services a customer needs (and in the correct sequence) to form a single complex service (cp. for instance Fig. 3.7). Other than AppExchange and TEXO, it builds upon the ideal conception of lightweight approaches. RESTful architectures and simplified interfaces have dramatically reduced the complexity of service composition and service description, finally enabling automated on-demand orchestration of complex services. This assumption is revisited in Sect. 3.5.

3.4.4.1 Payment Processing

Consider the mid-sized company TulipBuddy that distributes flowers over the Internet. As payment processing is not TulipBuddy's core competency of, its board decides on the integration of third-party Web services to boost existing business processes in order to decrease costs of operation and maintenance. Figure 3.6 sketches an excerpt of the service components of an exemplary complex service to provide payment processing functionality.

The *PaymentProcessingService* facilitates service components from StrikeIron, Duo Share,[16] and CDYNE[17] to verify customers' addresses and credit card information.

Customer data is managed and stored utilizing a DataBaseService and a StorageService, each provided by third parties. For example, services for organizing

[16]http://duoshare.com/.

[17]http://cdyne.com/.

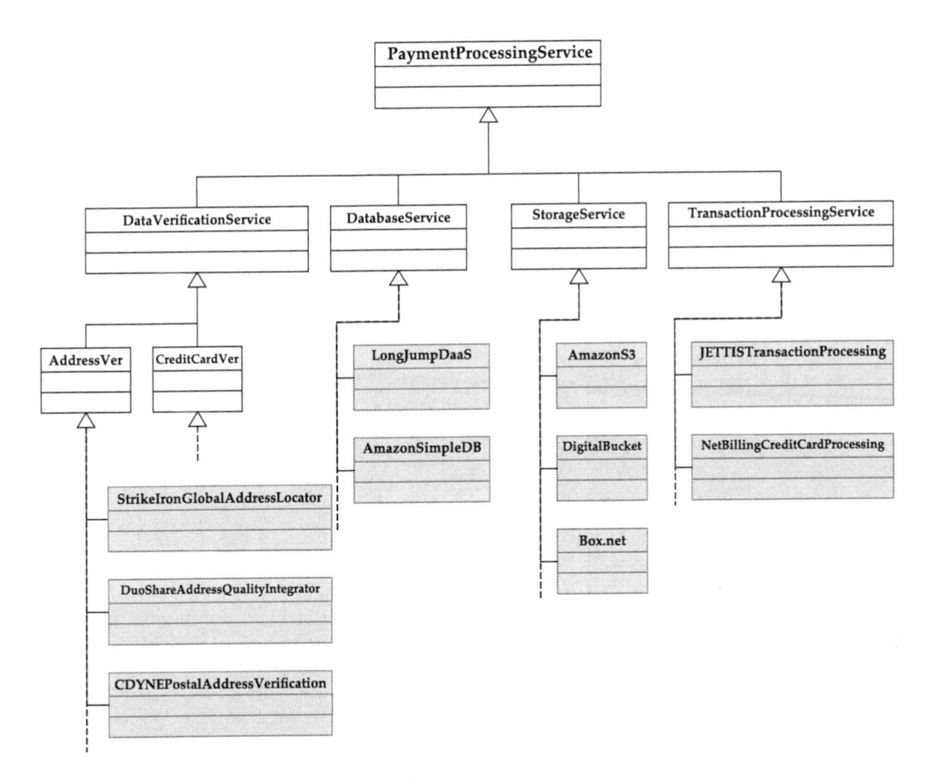

Fig. 3.6 Payment processing service (static view)

and managing customer data are Amazon Simple DB and Long Jump DaaS.[18] Exemplary third-party storage services are Amazon S3, Digital Bucket,[19] and Box.net.[20]

The execution of the financial transaction through the TransactionProcessing Service is, for instance, provided by JETTIS Transaction Processing[21] and Net Billing Credit Card Processing.[22]

Figure 3.7 shows the process behavior of the payment processing complex service. In a first step, customer data is validated. Thereafter, the actual transaction occurs (transaction processing service) which ends with and the charge of the customer's credit card account. The change in state must be updated consequently in the internal database of the company. The database service updates corresponding customer data that is stored using a decentralized storage service.

[18]http://longjump.com/daas/.

[19]http://digitalbucket.net/.

[20]http://box.net/.

[21]http://www.jettis.com/.

[22]http://www.netbilling.com/.

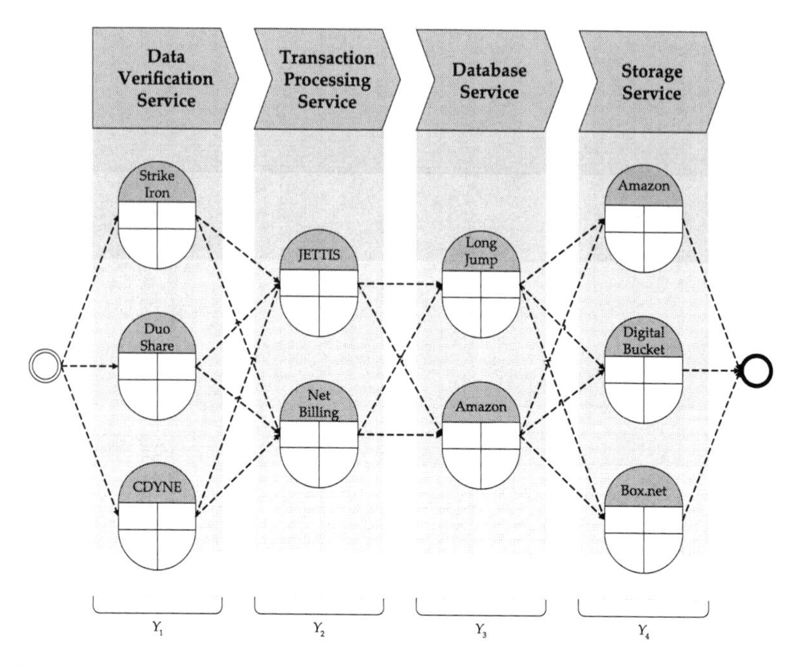

Fig. 3.7 Payment processing service (dynamic view)

For each process step, the candidate pool of the SVN yields potential suitable candidates to fulfill required business transaction. The result of each transaction is passed to the successor service. In order to successfully instantiate the complex service, the overall transaction requires exactly one service candidate from each pool.

3.4.4.2 MeetingAssist

This example depicts the formation of an SVN that is able and ready to instantiate a complex service based on the requirements imposed by service request. Assume that a customer that requires a complex service that scans all her calendar entries of the next week as to the upcoming meetings. Based on the meetings' descriptions, the complex service shall query the soft skills of all meeting participants consulting information that is stored in their social community profiles. The retrieved information is then updated in a CRM data base which is stored in an on-demand storage infrastructure.

A set of service providers participates in the SVN that offers the above-described MeetingAssist complex service, each grouped in candidate pools. Google offers its

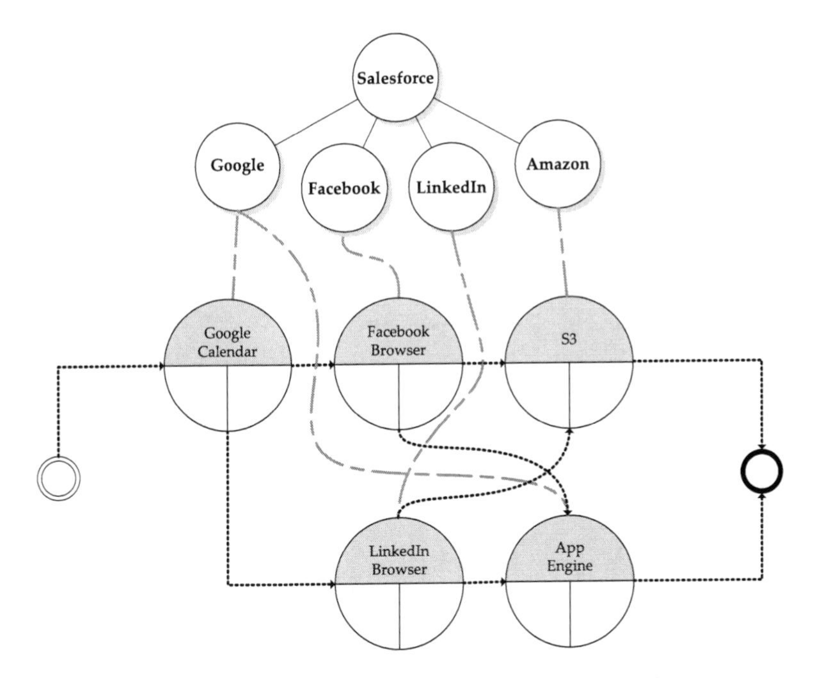

Fig. 3.8 Example of a service value network realizing a CRM complex service

Google Calendar service[23] and Google App Engine[24] which provides a scalable infrastructure for service development and storage. The social community platforms Facebook[25] and LinkedIn[26] provide services to browse profiles of registered users. The Simple Storage Service (S3) by Amazon offers flexible storage capabilities. As depicted in Fig. 3.8, in case of the underlying SVN example, the requested complex service can be realized in four different variants by selecting feasible service combinations (e.g., Google Calendar, Facebook Browser and Amazon S3).

3.5 Conclusion

The Web services domain has turned out be the perfect environment for SVNs. Specialization through service composition is unique to Web services. Web services bring about the possibility of automated on-demand composition to complex services and feature modularity as well as well-defined interfaces. Moreover, by their transmission via a Web protocol, provisioning of business value can take place

[23]http://google.com/calendar.

[24]http://code.google.com/appengine/.

[25]http://facebook.com/.

[26]http://linkedin.com/.

via a universally accessible network. Automated on-demand composition through a universally accessible orchestration platform is the distinguishing factor for SVNs, as shown in a comprehensive overview on related networked organization forms: in Sects. 3.2 and 3.3, a definition of SVNs and their distinction from related concepts as well as a theoretical formalization were provided. Both contributions are the very foundation for designing mechanisms to automatically allocate and price complex services (cp. Chap. 7).

Furthermore, the idea of SVNs is extended by a service intermediary that composes services for customers in Chap. 6, in order to enable the intermediary to set incentives for contributing service providers with respect to their level of effort for service execution.

The focus of this concluding section is twofold. Firstly, the benefits for participants in SVNs shall be clarified. Secondly, the minimum requirements as to standards that are imposed by the platform operator shall be revisited due to differences between the ideal conception and the prior art.

3.5.1 Benefits for SVN Participants

Earlier business applications provided over the Internet were designed to catch the average customer requirements, but failed to offer customization possibilities and left special requirements unsatisfied (Herbert and Martorelli 2008). A survey conducted by the Sandhill Group and McKinsey and Company showed that approximately 75% of the small and medium-sized enterprises (SME) polled would be ready to use some sort of on-demand software offerings (Dubey et al. 2008). However, SME seek for *tailored* solutions, too. In SVNs, the procurement of affordable, customized applications on demand is made possible by virtue of automated service composition.

Service providers within an SVN are able to exploit the benefits of the combinatoric and network effects. A single service offering can become a complementary part in multiple valuable complex services. That way, each single service will generate more value than it would as a stand-alone offering, and thus, it is more likely to be purchased.

Furthermore, a new hybrid role is fostered in SNVs – the *prosumers* (Tapscott and Williams 2006), both benefitting from the flexible service offerings as a consumer *and* contributing to the SVN's value as a provider. Services presently not available in the SVN and, thus, having been programmed by the customer itself, can be contributed to the SVN and have the potential to kiss awake latent demand.

3.5.2 Minimum Requirements Imposed by the Platform Operator

As stated in the introduction of this book, Web service mashups came to life with emerging Web 2.0 technologies and the renaissance of HTTP appreciation.

Typically, these service mashups as consumed today can still predominantly classified as consumer or data mashups. Consumer mashups combine data elements from diverse sources and hide them behind a simple user interface (UI). Similarly, data mashups combine data streams into a single data feed with a dedicated UI attached to it. Analyzing the top mashup tags at Programmableweb.org, it becomes obvious that the most popular mashups still belong to above-mentioned categories, with *mapping*, *photo*, *shopping*, *social*, and *video* being among the top five used tags to circumscribe the mashup offered.[27] Among the most popular mashups on ProgrammableWeb.org, one finds, for instance, a mashup to help consumers locate a Nintendo Wii,[28] a mashup that features live weather, forecasts, webcams, etc.,[29] or a mashup that grabs sad tweets from Twitter[30] and illustrates them via pictures taken from Flickr.[31]

These mashups mainly do base on RESTful Web services that encapsulate functionality and put them behind clearly defined interfaces based on HTTP, thereby facilitating lightweight approaches such as RESTful architectures and slim messaging formats. However, this is only partially true for business-oriented, enterprise mashups. While ProgrammableWeb.org increasingly lists APIs of business-related Web services that can potentially be used in mashups providing business application functionality,[32] other platforms rely on minimal requirements that are a bit higher. In respect of valuable and interoperable business applications, composition is not as simple as that (Weiss and Gangadharan 2009). Both salesforce.com's AppExchange and the TEXO platform impose *open, but proprietary standards*. As shown in Sects. 3.4.2 and 3.4.3, AppExchange requires services to be implemented according to Apex via Force.com; TEXO requires service providers to implement their service offerings using ISE Workbench. Both approaches enable seamless composition and compatibility of service modules from a technical point of view.

Indeed, such open but proprietary standards do not fully reflect the lightweight, ideal conception of SVNs. However, it is important to note that the approaches pursued by AppExchange and TEXO do not contradict the notion of SVNs. As long as the registration is open for any service that meets the defined minimum requirements, in particular with respect to appropriate interface specifications, proprietary standards are acceptable (cp. Definition 3.1).

---

[27] The data is taken from http://www.programmableweb.com/mashups, Status: 03/21/2011.

[28] http://www.programmableweb.com/mashup/wii-seeker.

[29] http://www.programmableweb.com/mashup/weather-bonk.

[30] http://twitter.com/.

[31] http://www.programmableweb.com/mashup/sad-statements.

[32] For instance, by April 2010, approximately 80 enterprise service APIs were listed on ProgrammableWeb.org offering functionality such as CRM or IT management. These Web services are potentially available to be included in a value-added enterprise mashup.

References

Araujo L, Spring M (2006) Services, products, and the institutional structure of production. Ind Market Manag 35(7):797–805

Baker W (1992) The network organization in theory and practice. In: Nohria N, Eccles R (eds) Networks and organizations: Structure, form, and action. Harvard Business School Press, Boston, pp 397–429

Baldwin C, Clark K (2000) Design Rules, Vol. 1: The power of modularity. MIT, Cambridge

Barros A, Dumas M (2006) The rise of web service ecosystems. IT Prof 8(5):31–37

Basole R, Rouse W (2008) Complexity of service value networks: Conceptualization and empirical investigation. IBM Syst J 47(1):53–70

Bengtsson M, Kock S (2000) "coopetition" in business networks – to cooperate and compete simultaneously. Ind Market Manag 29(5):411–426

Bitsaki M, Danylevych O, Heuvel W, Koutras G, Leymann F, Mancioppi M, Nikolaou C, Papazoglou M (2008) An architecture for managing the lifecycle of business goals for partners in a service network. In: Proceedings of the 1st European conference on towards a service-based internet, Madrid, pp 196–207

Busquets J, Rodon J, Wareham J (2009) Adaptability in smart business networks: An exploratory case in the insurance industry. Decis Support Syst 47(4):287–296

Chesbrough H, Spohrer J (2006) A research manifesto for services science. Commun ACM 49(7):35–40

Davidow W, Malone M (1993) The virtual corporation: Structuring and revitalizing the corporation for the 21st century. Harper Business, New York

Devlin G, Bleackley M (1988) Strategic alliances – guidelines for success. Long Range Planning, New York 21(5):18–23

Downes L, Mui C (2000) Unleashing the killer App: Digital strategies for market dominance. Harvard Business School Press, Boston

Dubey A, Wagle D (2007) Delivering software as a service. Online, URL http://www.mckinsey.de/downloads/publikation/mck_on_bt/2007/mobt_12_Delivering_Software_as_a_Service.pdf, the McKinsey Quarterly, Web exclusive

Dubey A, Mohiuddin J, Baijal A, Rangaswami M (2008) Enterprise software customer survey. Sand Hill Group, McKinsey and Company http://www.interop.com/downloads/mckinsey_interop_survey.pdf

Goldman S, Nagel R, Preiss K (1995) Agile competitors and virtual organizations: Strategies for enriching the customer. Van Nostrand Reinhold, New York

Hakansson H, Ford D (2002) How should companies interact in business networks? J Bus Res 55(2):133–139

Hamel G, Doz Y, Prahalad C (1989) Collaborate with your competitors – and win. Harv Bus Rev 67(1):133–139

Hamilton J (2004) Service value networks: Value, performance and strategy for the services industry. J Syst Sci Syst Eng 13(4):469–489

Hamilton J (2007) Service value networks: Into practice. J Syst Sci Syst Eng 16(4):414–423

Harel D, Naamad A (1996) The statemate semantics of statecharts. ACM Trans Software Eng Meth 5(4):293–333

van Heck E, Vervest P (2007) Smart business networks: How the network wins. Commun ACM 50(6):28–37

van Heck E, Vervest P (2009) Smart business networks: Concepts and empirical evidence. Decis Support Syst 47(4):275–276

Herbert L, Martorelli B (2008) SaaS clients face growing complexity. Forrester Study

Hinterhuber A (2002) Value chain orchestration in action and the case of the global agrochemical industry. Long Range Planning 35(6):615–635

Hoetker G (2006) Do modular products lead to modular organizations? Strat Manag J 27(6): 501–518

Holland C, Lockett A (1998) Business trust and the formation of virtual organizations. In: Proceedings of the 31st Hawaii international conference on system sciences (HICCS), Honolulu, vol 6, p 602–610

Holm D, Eriksson K, Johanson J (1996) Business networks and cooperation in international business relationships. J Int Bus Stud 27(4):1033–1053

Hoogeweegen M, Vervest P (2005) How much business modularity? In: Vervest P, van Heck E, Preiss K, Pau L (eds) Smart Business Networks. Springer, Heidelberg, chap 23

Jones C, Hesterly W, Borgatti S (1997) A general theory of network governance: Exchange conditions and social mechanisms. Acad Manag Rev 22(4):911–945

Kabzeva A, Hillenbrand M, Müller P, Steinmetz R (2009) Towards an architecture for the internet of services. In: Proceedings of the 35th Euromicro SEAA conference, Patras

Miles R, Snow C (1986) Organizations: New concepts for new forms. Calif Manag Rev 28(3): 62–74

Mowery D, Oxley J, Silverman B (1996) Strategic alliances and interfirm knowledge transfer. Strat Manag J 17:77–91

Papazoglou M (2007) Web services: Principles and technologies. Prentice Hall, New Jersey

Ritter T, Wilkinson I, Johnston W (2004) Managing in complex business networks. Ind Market Manag 33(3):175–183

Scheithauer G, Voigt K, Bicer V, Heinrich M, Strunk A, Winkler M (2009) Integrated service engineering workbench: Service engineering for digital ecosystems. In: Proceedings of the international conference on management of emergent digital ecosystems (MEDES), Lyon, pp 446–449

Spohrer J, Anderson L, Pass N, Ager T, Gruhl D (2008) Service science. J Grid Comput 6(3): 313–324

Stathel S, Finzen J, Riedl C, May N (2008) Service innovation in business value networks. In: Proceedings of XVIII international RESER conference, Stuttgart, pp 288–302

Strader T, Lin F, Shaw M (1998) Information structure for electronic virtual organization management. Decis Support Syst 23(1):75–94

Tapscott D, Williams A (2006) Wikinomics: How mass collaboration changes everything. Portfolio, New York

Tapscott D, Ticoll D, Lowy A (2001) Digital capital: Harnessing the power of business webs. Harvard Business School Press, Boston

Thorelli H (1986) Networks: Between markets and hierarchies. Strat Manag J 7(1):37–51

Tiwana A (2008) Does technological modularity substitute for control? A study of alliance performance in software outsourcing. Strat Manag J 29(7):769–780

Vargo S, Lusch R (2004) Evolving to a new dominant logic for marketing. J Market 68(1):1–17

Vervest P, Preiss K, Van Heck E, Pau L (2004) The emergence of smart business networks. J Inf Technol 19(4):228–233

Vervest P, van Heck E, Preiss K, Pau L (2005) Smart business networks. Springer, Berlin

Weiss M, Gangadharan G (2009) Modeling the mashup ecosystem: Structure and growth. R&D Manag 40(1):40–49

Williamson O (1973) Markets and hierarchies: Some elementary considerations. Am Econ Rev 63(2):316–325

Wolters M, Hoogeweegen M (1999) Management support for globally operating virtual organizations: The case of klm distribution. In: Proceedings of the 32nd annual Hawaii international conference on system sciences (HICCS), Maui, p 7015

Yang J (2003) Web service componentization. Commun ACM 46(10):35–40

Yoshino M, Rangan U (1995) Strategic alliances: An entrepreneurial approach to globalization. Harvard Business School Press, Boston

Chapter 4
Business Models

Business models are oftentimes considered as "the most discussed and least understood aspect of the web. There is so much talk about how the web changes traditional business models. But there is little clear-cut evidence of exactly what this means" (Rappa 2001). Today, a whole decade later, this statement is still very true, especially when is is transferred to business models in newly arising business networks that are, for instance, fostered by Web service technologies. Business networks arise in an environment of intense but possibly short-run cooperation between different service providers and require a high degree of complementarity of services. For these requirements to be met, on the one hand technological changes need to take place. On the other hand, business networks for the automatic provisioning of Web services have some special economic features which must find consideration in the respective business models before they can be realized.

Despite a rough understanding of business models that seems to be widespread – namely an aggregation of essential, relevant aspects from economic branches providing a compressed overview on business activities – dimensions, components, perspectives, and core issues are depicted non-uniformly and confusingly (Petrovic et al. 2001; Wirtz 2001). In particular, current literature does not provide a comprehensive overview on the work that has been done so far. What is more, the state-of-the-art business model understanding, itself being not yet fully grasped, cannot be mapped "as is" to business models of companies providing services in business networks.

Moreover, a business network, expecially in the context of automatic service provisioning, can be seen as a conglomerate of each participating provider's network of partners, as each provider does only know about their direct partners. Even though more and more companies enter the field of automatic service provisioning and composition, academia lacks contributions that particularly consider business model frameworks for such networked economies.

In order to address these shortcomings, this chapter introduces a business model framework for business networks, and for service value networks, in particular. This business model framework is a result from an extensive literature review comprising two areas: On the one hand, a state-of-the-art analysis of existent definitions of

C. Weinhardt et al., *Business Aspects of Web Services*,
DOI 10.1007/978-3-642-22447-8_4, © Springer-Verlag Berlin Heidelberg 2011

the term *business model* with emphasis on the elements that should be included, is conducted. In this connection, a segregation of the concept business model from related concepts is also given. On the other hand, in order to establish a business model framework for business networks, the characteristics of business networks as presented in Sect. 3 are reviewed.

Thus, this chapter shall establish a common understanding of the concept of business models within business networks. It is structured as follows: Sect. 4.1 gives an introductory overview of the concept "business model," pin-pointing the origin of the term and the different usages of the concept. In Sect. 4.2, the extensive body of academic literature on business models, incorporating 26 state-of-the-art business model definitions, is discussed. Based on these considerations, the term "business model" is differentiated from related concepts. A complete overview on business model taxonomies rounds off this section. Taking these insights as one building block for the definition of a business model framework and the results from Chap. 3 as the second module, an additional element of the latter is discussed in Sect. 4.3: customer-centricity. As a symbiosis of the results provided in Chap. 3, Sects. 4.2 and 4.3, the *business model framework for business networks* is presented in Sect. 4.4.

Thus, the contribution of this chapter is as follows: Providing the business model framework, an updated understanding of the concept of business models within business networks will be established. It is the basis for further research, be it for the creation of a business model taxonomy in networked economies, the analysis of existent networked business models, or the development of concrete business models in a business network context.

Especially the latter application of the business model framework in this chapter is a major factor of success for providers of services and for composition platform operators. The business model framework that is tailored to their needs will identify the business aspects of Web services and will support important design decisions as to the implementation of a particular business model.

The following chapters will mainly focus on one part of the business model framework, that is, the revenue model, by means of pricing decisions. For example, Sect. 7 illustrates an approach for price determination, whereas Sect. 6 takes strategic decisions by means of incentive design into account.

4.1 The Concept "Business Model"

Before deep-diving into the analysis of state-of-the-art business models, this section provides a brief survey of the origin of the very concept "business model" and its development. Thereafter, different perspectives on the term "business model" as well as different understandings of what a business model actually is, and which elements it is comprised of, respectively, are outlined and distinguished.

4.1.1 *Origin and Development of the Concept "Business Model"*

The term "business model" is being intensively discussed in literature since the mid 1990s. Osterwalder et al. (2005) used a method by Abrahamson and Fairchild (1999) to trace the appearance of "business model" in journals.[1] The query confirms that, even though appearing in 1957 the first time,[2] research gathered momentum not until the mid 1990s. At about the same the first Internet-based e-commerce solution emerged. In the course of the Internet boom (1999–2001), the concept "business model" has once again gained increased popularity.

Surprisingly, a consistent definition of what a business model is comprised of did not win recognition (Sandrock and Weinhardt 2006), even though a couple of authors are repeatedly cited (e.g., (Timmers 1998, 1999; Alt and Zimmermann 2001; Chesbrough and Rosenbloom 2002; Wirtz 2001; Stähler 2002) in the German-speaking community). Timmers was the first author to present a definition that gained acceptance though never prevailed as the one and only notion: A business model is the organization (or "architecture," as Timmers calls it) of product, service and information flows, and the potential benefits and sources of revenues for involved (business) actors (Timmers 1999, p. 32). In addition, there is not only dissent in the essence of a business model – there is also dissonance in the very utilization of the term "business model."

Determining the concept of a business model solely by means of an analysis of the term is suitable for a first introduction, but certainly the outcome is altogether far too sketchy (Schögel 2002). A "model" is a simplified representation of a fact from reality. Transferred to a business environment, the outcome of such a representation should be a structure-alike or at least homomorphous mapping of a business activity. Derived from the general concept of "model," the purpose of a business model is to represent the elements and relations of a business (activity) in such a simplified way that the mostly highly complex real system can be wholly captured (Bamberg and Coenenberg 2002, p. 13). Conversely, it must be understood that a model is neither able nor willing to provide an entirely adequate and consistent representation of reality – a model is therefore subject to "focused attention" (Petrovic et al. 2001) while detracting from other factors. A total capture of the business activities would be inappropriate or by unreasonable means and efforts. Whereas the understanding of the term "model" is relatively clear, the scope of the term "business" cannot be determined intuitively and depends on the range one assigns to the essence of a business. Such a definition requires a specification of the minimal set of elements and properties necessary to map a considered business into a model.

Since the beginning of the twenty first century, the basic parameters for e-business have changed. The burst of the dot com bubble had the new economy held

[1] Titles, abstracts, keywords, and full texts of all articles in the Business Source Premier database of scholarly business journals were searched electronically for the string "business model."

[2] Bellman et al. (1957) used the term in an article published in a journal for operations research.

its breath. As a result of the crash, countless Internet start-ups have become insolvent with their investors losing both huge amounts of money and their confidence in new markets. Contrary to common belief, O'Reilly[3] was convinced that rather than being damaged by the market crash, the Web was now even more important than ever before (O'Reilly 2005). This milestone was subsequently referred to as the beginning of Web 2.0 – the selection process resulting from the burst of the bubble filtered those businesses that had generated sustainable and convincing concepts. Those concepts were in their infancy in the early 2000s, today they are becoming more and more pre-requisite for companies that wish to be ready for an environment moving towards new paradigms such as customer empowerment and integration, self-organization and peering, mass collaboration and the likes (Tapscott and Williams 2006). Even more important, the way value is created in networks has changed. While outsourcing succeeded vertical integration and launched the trend to a joint value creation process in the early 1990s, business networks are on their way to transform themselves to more and more loosely-coupled networks which heavily impact on the business models of participating partners (cp. Chap. 3).

Noticeably, new contributions to business model research incorporating the paradigm changes mentioned above are scarce. Why so? Is the step from the first wave of e-business to Web 2.0-related business too small for the development of new business model concepts? From a mere technological perspective, one could support this thesis. Osterwalder (2004, p. 12) emphasizes that major progress in information and communication technology (ICT) has made it possible to offer entirely new products and services, or traditional products and services via a whole new distribution channel, respectively. Thanks to the Internet, customers can be approached in a new and innovative way, radically pushing away physical restrictions. This was a major cut, in fact longing for whole new strategies and concepts to approach the newly arising electronic markets with a multitude of new pricing and revenue mechanisms opening up. With respect to this far-reaching impact of the world wide web, the change from first- to second-generation e-business is naturally smaller than the very possibility to conduct business via the Internet.

ICT has also significantly reduced transaction and coordination costs. Today, the benefit of integrating partners and customers in inter-company processes and communication networks has exceeded its costs many times over (Pohle et al. 2005). Such cost reductions can be interpreted as fundamental drivers towards collaboration in business networks and active customer integration, the latter being postulated a vital element of Web 2.0. Certainly, it is not only ICT that pushed the customer from the mere buyer status right into the center of consideration, it is also the shift in the customers' demands themselves towards customized, holistic solutions. This shift is closely related to the emergence of networks since the more complicated products and services get, the more specialized partners are required to

[3]Tim O'Reilly is founder and CEO of O'Reilly Media. He is reckoned a pioneer supporter of the open source movement and is widely credited with coining the term Web 2.0, http://en.wikipedia.org/wiki/Tim_O'Reilly – 2011-03-01.

contribute complementary elements to an integrated overall product or service. Not only the isolated product or service, but the whole ecosystem of suppliers needs to be considered (Kagermann and sterle 2006, p. 15).

4.1.2 *Different Perspectives and Usage of the Term "Business Model"*

Academic literature does not utilize and understand the term "business model" the same way. Linder and Cantrell point out that business models can usually be interpreted in three different ways (Linder 2000): As actual components of business models, real operating business models, and change models. Another field which is oftentimes addressed when referring to business models in a more formal sense is an overarching concept of business models (Osterwalder et al. 2005), in terms of a systematizing reference model (Alt and Zimmermann 2001). The common understanding of "business model" from generic to concrete can be classified in four categories:

- Overarching concept in terms of a systematizing reference model
- Distinctive parts of business models (e.g., the revenue model)
- Types or typologies of business models (taxonomies)
- Concrete instances of business models

To clarify the different categories mentioned above, an illustration based on Osterwalder et al. (2005) is consulted. This categorization shown in Fig. 4.1 will be used in Chap. 4.2 to classify existent contributions to business model science.

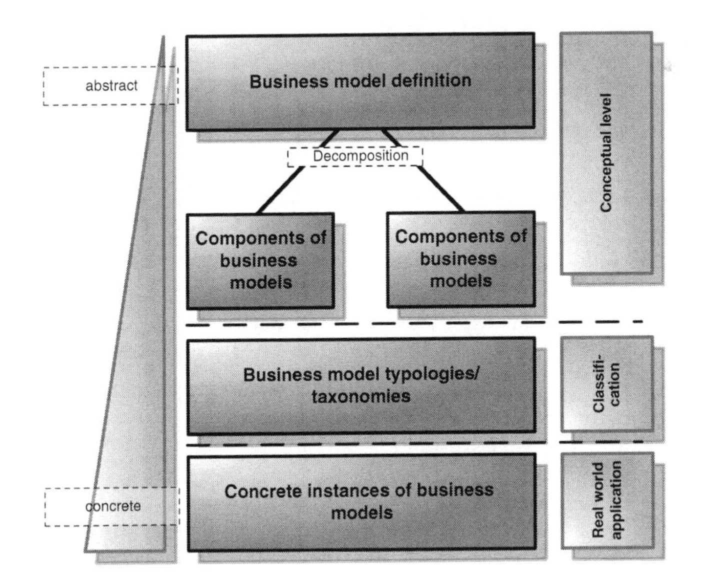

Fig. 4.1 Distinction of different business model understandings

(a) *Conceptual level:* In this abstraction level, definitions of what a business model actually is and of which components/elements a business model is generally comprised can be found. The latter can be seen as a meta-model (Osterwalder et al. 2005). Single elements of a business model like the revenue model are indeed parts of the whole, but as themselves not yet a business model. Occasionally, specific types of pricing and revenue mechanisms are misleadingly titled a "business model," especially in more practice-oriented coherencies. Nevertheless, a business model is more than having a pricing scheme (Linder 2000; Osterwalder et al. 2005). This issue is elaborated in Sects. 4.2 and 4.4.
(b) *Classification:* Taxonomies can be defined as a classification of business model types into categories according to common characteristics. The basis for such a typology can be manifold. A categorization strategy that is often utilized is based on specific elements of business models. Wirtz (2001) applies classifications based upon terms of the separation criteria "value creation" whereas other authors (e.g., (Betz 2002; Eisenmann et al. 2001; Rappa 2001; Timmers 1999; Weill and Vitale 2001)) use categories that use more than one criterion. Note that it is usually not possible to distinctly separate different groups – overlaps of the classifiers have to be accepted (Wirtz 2001, p. 87).
(c) *Real-world application:* In this case, the term "business model" denotes concrete existent or planned real-world business models, which are in scientific purpose oftentimes consulted and described when analyzing and/or comparing companies (Chesbrough and Rosenbloom 2002).

From the above-mentioned different approaches to business model concepts, different intentions of the usage of business model definitions can be observed:

- As a basis for classifier for business models
- As an analyzer and description unit for present real-world business models or
- As an advisor for the implementation of concrete business models on a strategic level (Alt and Zimmermann 2001)

Lastly, a differentiation according to the background of the authors seems to be a fruitful approach to understand the origins of different perspectives and accentuations of the concept. While some authors' background is mainly based on strategic considerations (cp., e.g., (Magretta 2002; Mller-Stewens and Lechner 2001)), others approach the topic from a traditional economic perspective focussed on e-business (cp. e.g., (Stähler 2002; Timmers 1998; Wirtz 2001; Yu 2001)), and a whole range of contributions is dedicated to implementing models or ontologies after formally defining the concept "business model" (cp., e.g., (Lindström 1999; Nilsson et al. 1999; Osterwalder and Pigneur 2002; Osterwalder 2004)).

4.2 State-of-the-Art Business Models: An Analysis

After introducing the concept of business models and showing different perspectives on how the term can be interpreted in Chap. 4.1, existing state-of the art definitions and frameworks for business models are elaborated in this section.

4.2.1 Methodology

Subject to study shall be preferably independent business model definitions put forth by academic literature in recent years. Definitions and frameworks that are based on earlier work are only included if they considerably modify the piece of work they are built upon.

The general sampling criterion for the choice of sources was the amount of relevant citations in related literature as well as mentions in literature that is also concerned with giving an overview on business model definitions and typologies such as Bieger et al. (2002), Osterwalder (2004), Sandrock and Weinhardt (2006), Scheer et al. (2003), Schwickert (2004), Stähler (2002). Objective of this overview is to extract essential elements of business models from state of the art definitions.

To classify these definitions, they are assigned to the scheme presented in Fig. 4.1. Further, the disciplinary background of the authors is given (cp. Fig. 4.5). The result will be a consolidated view on existing definitions, frameworks, and classification schemes based on a qualitative and quantitative analysis. Using this result, the concept "business model" can be differentiated from related concepts that are oftentimes wrongly utilized as synonyms.

4.2.2 Overview on State-of-the-Art Business Models: A Qualitative Analysis

As stated above, business models in newly arising business networks are poorly supported by scientific research. Additionally, state-of-the-art business models are themselves not yet fully grasped and far from consistent. Before identifying the requirements that business networks, and service value networks in particular, impose for business models in this context, the state-of-the-art landscape of business model research is to be scrutinized. On the one hand, the objective is to elaborate elements that are commonly supported in literature. On the other hand, attention is turned upon elements that are subject to controversy.

Having in mind the context of business networks, it is needless to say that particular emphasis is put upon elements associated with joint value creation and core competencies of partners, modularization, flexibility, and the likes. It is also worth mentioning that the German-speaking community accounts for a considerable amount of business model literature which is translated and presented to the reading public of this book.

4.2.2.1 Timmers (1998, 1999)

Timmers (1998, 1999) exclusively relates the concept "business model" to Internet-based e-commerce. He refers to the inconsistencies in the usage of the term, yielding the following definition: A business model is

- "An architecture for product, service and information flows, including a description of the various business actors and their roles; and
- A description of the potential benefits for the various business actors; and
- A description of the sources of revenue." (Timmers 1999, p. 32)

The above-mentioned architecture is built from the interplay of the actors ("interaction patterns" (Timmers 1999, p. 33)) and the integration of information into the value chain. The technical implementation of such an architecture depends on the currently available technology. Beyond business models, Timmers identifies a marketing model, which is composed of the business model in addition to the corresponding marketing strategy. Consequently, the marketing model is not part of a business model, it rather supplements it. Especially the competitive advantage – the following definitions show that authors heavily disagree upon this particular issue – is not an integral part of the business model according to Timmers. The competitive advantage can be interpreted as a direct consequence of a company's business model and its marketing model.

4.2.2.2 Lindström (1999)

Lindström (1999) also focuses on the process and purpose of "business modeling," not on the elements of a business model. Regarding to Lindström, business modeling deals with the two disciplines "business" and "information." The former is defined as using "a limited set of resources intended to create/increase customer-perceived value" (Lindström 1999, p. 152). The latter are eventually human thoughts or mental models that can be formalized via business modeling and thus made transparent to all stakeholders within an organization. Lindström mentions some modeling tools, which implicitly point at some possible elements of business models:

- Organigrams pinpoint the responsibilities within an organization
- Business process models describe actors, processes, and their interplay to attain maximum customer satisfaction

In sum, this definition is vague and non-specific with respect to elements of business models.

4.2.2.3 Nilsson et al. (1999)

The definition provided by Nilsson et al. (1999) was written as an introductory chapter for a book on business modeling (Nilsson et al. 1999, p. 1). Due to the technical background of the authors, they emphasize the use of models and methods to both understand and change organizations. They identify the bridging of information gaps between business people and system people to be main objective of modeling a business. Thus, most importantly, this contribution to business model literature does not define the business model itself or the elements subsumed under the business activity of an organization, but the concept of "business modeling." For the purposes in this book, the definition is applicable only in a limited degree since the focal point is the development ("modeling") process itself rather than the intrinsic content of a business model.

4.2.2.4 Hamel (2000)

Hamel (2002) identifies and describes four components of a "business concept," which is, according to the author, equated with the term "business model." The author subdivides the business concept into several elements:

- Core strategy
  - *Business mission:* General objective of the strategy, e.g., value proposition, purpose, high level aims, etc.
  - *Product/market scope:* Positioning of the company
  - Basis for differentiation
- Strategic resources
  - Core competencies
  - Strategic assets such as brands, patents, infrastructure, and standards
  - *Core processes:* Value creation processes transforming input factors into output, thus generating value for the customers. Herein, Hamel identifies a fundamental starting point for innovation (Hamel 2002, p.80)
- Customer interface
  - Fulfillment and support describes the go-to-market strategy, distribution channels and customer relationship management
  - Information and insight about customers
  - Relationship dynamics in respect to the customer (e.g., direct, indirect, firm, etc.)
  - Pricing structure, including pricing and revenue models
- Value network
  - Suppliers
  - Partners, who typically generate complementary goods and services contributing to the overall product or solution, respectively
  - Coalitions

These components are interconnected via three "bridges" (Hamel 2002, p. 73):

- "Customer benefits" link the core strategy with the customer interface and highlights the benefits for the customer
- "Configuration" interconnects core strategy with the strategic resources, merging competencies, assets, and processes
- "Company boundaries" link the strategic resources and the value network, modeling competencies and relationships between actors

Hamel depicts a very extensive definition of a business model, integrating both the concepts of strategy and marketing. Hence, Hamel disassociates himself from other contributions to literature with respect to the scope of the term.

4.2.2.5 Heinrich (2000), Heinrich and Leist (2000)

Heinrich (2000) and Heinrich and Leist (2000) introduce a business model framework for commercial banks in the era of e-business. According to the authors, a business model is a description of the actual condition of a company in terms of market, value creation, and potential. To describe such a condition representative, endogenous, and determinable dimensions have to be found. These dimensions can be divided into an external view – market appearance of a company – and an internal structural view.

- The external view considers the market-related business configurations and their interdependencies, including target markets, market and customer segments, relation to the customers (in terms of value added, compensation, and communication). Furthermore, the value creation, its logic, and its success factors, as well as distribution channels are considered. The definition is strongly marketing-driven.
- The internal view mainly characterizes cooperation partners, degree of coordination between distribution channels, and locations. Furthermore, the organizational structure and competencies are considered.

The business model should represent the essential results of the strategy planning. Hence, it needs to be differentiated from the strategy.

4.2.2.6 Klueber (2000)

Klueber (2000) defines business models as "summary of the value creation logic of an organization or a business network including assumptions about its partners, competitors and customers." A business model defines "the business and IS architecture, rules, potential benefits and sources of revenue" (Klueber 2000, p.797). Interestingly, Klueber highlights that his definition is also valid for business networks. (Klueber 2000) includes competitors as well as technology as integral parts of business models in contrast to other authors who see these elements as external forces impacting on business models.

4.2.2.7 Mahadevan (2000)

Mahadevan (2000) accentuates three fundamental aspects of a business model:

- Value streams between business partners and customers which identifies the value proposition for the buyers
- The revenue stream describes how revenue is generated and assured and
- The logistical stream "addresses various issues related to the design of the supply chain for the business" (Mahadevan 2000)

Mahadevan uses each of these streams to define a more detailed topology classifying business models refining the pre-identified three basic market models "portal," "product/service provider," and "market maker" (Mahadevan 2000).

4.2.2.8 Zimmermann (2000)

Zimmermann (2000) bases his definition on the work provided by Timmers (1998, 1999), though extending it in several issues. "A business model is defined as [...]: An architecture for the product or service addressing certain customer needs, A definition of the relevant business community, including a description of the various agents and their roles and protocols of interaction, A description of the potential benefits for the agents, A description of the sources of revenue" (Zimmermann 2000, p. 729). In contrast to Timmers (1998, 1999), Zimmermann emphasizes the interplay and the pattern of action of the actors.

4.2.2.9 Bartelt and Lamersdorf (2000), Bartelt et al. (2001)

Bartelt and Lamersdorf (2000) consider business models for e-business, while the follow-up paper is especially dedicated to e-information (Bartelt et al. 2001). The definition of business models is based on Timmers (1999), being an architecture consisting of products, services, information flows, and a description of involved actors. Additionally, the benefits for the actors and the sources of revenue should be accentuated (Bartelt et al. 2001). The major contribution of the two papers are classification schemes for e-business models (Bartelt and Lamersdorf 2000), and a classification scheme for e-information models (Bartelt et al. 2001). In the former, the classification is made by means of "economic agent" (provider, intermediary, consumer) and "type of communication" (active, passive), the latter is categorized by the domain of the e-information business model, namely literature, internet, and business-related information.

4.2.2.10 Afuah and Tucci (2001)

A business model is defined to be the method which mainly addresses:

- How a firm creates value, i.e., how resources are built and used within the value creation process. In this consideration, the target segment of customers and which product/service to offer to which segment are important elements of a business model.
- The money making potential that lays behind above-mentioned value proposition. Afuah and Tucci consider short-term as well as long-term revenue.

Thus, a business model is a system built upon components, linkages between components, and dynamics. Afuah and Tucci name ten components of a business model (Afuah and Tucci 2003, p. 51):

- *Profit site:* Relative advantage of a firm vis-á-vis “its suppliers, customers, rivals, potential new entrants, complementors, and substitutes” (Afuah and Tucci 2003, p. 53).
- *Customer value:* Differentiation of the offered product from competitor’s products. This can be done in several dimensions (product features, timing, location, service, product mix, linkage between functions, linkage with other firms, and reputation).
- *Scope:* Allocation of offered products to market segments (target customer).
- *Price:* Pricing is considered separated from revenues, the component describes the pricing strategy of a firm, e.g., menu pricing, one-to-one-bargaining, auction pricing, or barter.[4]
- *Revenue sources:* The sources of a firm’s revenue and profits are considered a crucial element of a business model. This includes financial flows indicating who pays for what and when.
- *Connected activities:* In the process of delivering value to the customer, firms have to perform additional activities such as research and development, product design, testing, marketing, sales, or field support. So, a business model has to clarify which activities in the value chain are performed and when.
- *Implementation:* The component “implementation” describes the transformation of reached business decisions, incorporating necessary organizational structure, systems, and people.
- *Capabilities:* Resources needed to create value. These assets can be tangible (physical and financial), intangible (e.g., patents or customer relationships), and human (skills and knowledge). (Core) Competencies and missing capabilities as well as the question how to fill such competence gaps are also included.
- *Sustainability:* The component sustainability is mainly a strategic one, outlining how a firm sustains its competitive advantage and thus keeps making money.
- *Cost structure:* Cost drivers in each component of the business model. Cost drivers can be economies of scale, technology costs, or transaction costs.

---

[4] Cp. also Sect. 4.4.2.1 as well as Chaps. 6–8.

Dynamics describe the changeability of a business model. A Business model has to be always ready to be reinvented due to environmental changes before competitors do.

According to the authors, a business model is the very cause for competitive advantage and thus, crucial to sustain a firm's profitability (Afuah and Tucci 2003, p. 73). Nevertheless, competitive advantage itself as well as related factors such as sustainability are considered an integral part of the business model. On the other hand, Afuah and Tucci depict technological change, and the environment (which, of course, can also be a competitors) of a firm as influencing factors which urge firms to innovate their business models. So, conversely, components dealing with the competitive advantage and sustainability of a business model as well as implementation-related elements can be considered an outcome or result of a business model rather than a component itself.

4.2.2.11 Alt and Zimmermann (2001)

Alt and Zimmermann (2001) utilize a literature review in order to identify the common denominator of all definitions. Doing so, they identify six generic elements of a business model (Alt and Zimmermann 2001):

- The mission includes a high level understanding of the vision pursued by the company, its strategic goals, and a value proposition pointing out the fundamental characteristics of the product or service.
- The structure determines the roles and agents which form a so-called business community, collaborating in a value chain or a value web. Moreover, the structure defines the focus on customer, industry, and products.
- The processes deliver detailed insight into the mission and structure, providing the elements of value creation process.
- Revenues are being highlighted as the bottom line of a business model. They include sources as well as necessary investments.
- Legal issues are being introduced as an element of business model, though rather being a limiting or shaping element for business models which impacts on each of the four above-mentioned elements (cp. Fig. 4.2).
- Technology takes over a double role, being enabler and constraint for business models at the same time. However, as well as legal elements, technology is not considered an integral part, but a shaping element for business models.

4.2.2.12 Amit and Zott (2001)

Amit and Zott (2001) provide a transaction-based definition of business models. Added value is grounded on the content, the structure, and the governance of the transactions (Amit and Zott 2001, p. 511). The "transaction content" describes the information and goods that are being exchanged as well as the resources and capabilities required to enable exchanges, that is auxiliary means. The "transaction

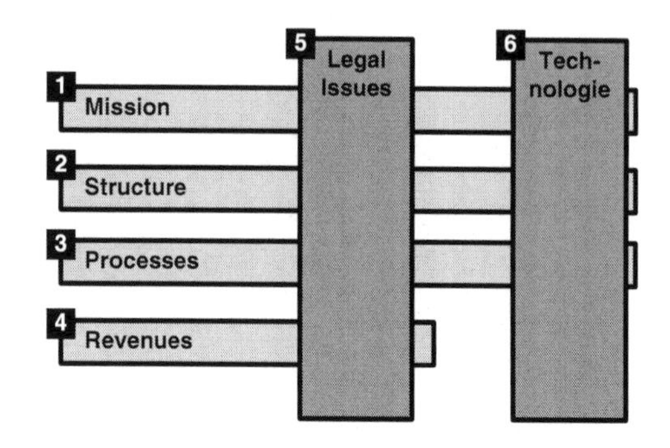

Fig. 4.2 Generic elements of a business model according to Alt and Zimmermann (2001). Source: Adapted from Alt and Zimmermann (2001)

structure" addresses the way actors are linked with each other when executing transactions. Moreover, the sequencing of transactions (i.e., the value creation process) and the mechanism of how value is created are considered. The "transaction governance" deals with actors in charge of monitoring and controlling flows of information, of goods, and of resources. Interestingly, the authors also include incentives for transaction participants into the business model definition. Being very closely linked to the transaction cost theory, the definition provided by Amit and Zott differs from other definitions in terms of not considering the output side (e.g., customers, distribution channel, revenue model, etc.).

4.2.2.13 Buchholz and Bach (2001), Bach et al. (2003)

According to the authors, the processes represented in a business model have to be targeted on the provision of value/benefit and enable revenue generation. Thus, a business model conduces to the implementation of competitive strategies instead of strategy being an integral part of the business model (Bach et al. 2003, p. 10). The authors identify four constituent components of business models (Buchholz and Bach 2001, p. 7) which are valid both for firms and for networks (Bach et al. 2003, p. 11): Process model, transaction model, participant model, and revenue model.

The process model describes the value creation processes. In case a single company is considered, its position in the value system is part of the description. The participant model outlines who contributes in which role to the value creation process. Since the customer is more and more incorporated into the value creation processes, he/she has become an increasingly important partner, not only due the trend towards customizing, but also by reason of the number of customers relevant to exploit network effects. Moreover, the participant model is affected by the coordination form of the value network. The transaction model answers the question how buyers and sellers come together (static vs. dynamic transaction modes). This partial model is interwoven with the forth pillar, the revenue model which outlines the logic of revenue generation and provides a scheme distributing these generated revenues within the network or the participating partners, respectively.

The definition provided by Bach et al. (2003) and Buchholz and Bach (2001) is universally applicable to individual firms and to networks, emphasizing the importance of a partner network which cannot be distinctly separated from the customer network anymore.

4.2.2.14 Rappa (2001)

Rappa sees business models as the "method of doing business by which a company can sustain itself" (Rappa 2001). For Rappa, the business model delineates how a company generates revenue by specifying its position in the value chain. Rappa's objective is not to define business models and its elements, he rather shows a taxonomy of business models which are present on the world wide web. These model types can be best characterized as a mixture of describing the service provided and the way revenue is generated: Brokerage Model, Advertising Model, Infomediary Model, Merchant Model, Manufacturer Model, Affiliate Model, Community Model, Subscription Model, and Utility Model.

4.2.2.15 Gordijn and Akkermans (2001)

Alike, the bulk of the literature considered, Gordijn and Akkermans (2001) concentrate on e-business models. Since the core purpose of their article is the introduction of ontology-based operators for e-business model de-construction and re-construction, the concept of a business model is introduced as a secondary contribution. An e-business model shows how a "network of actors creates, exchanges and consumes objects of value by performing value-adding activities" (Gordijn and Akkermans 2001). Though very short, the definition accentuates two important issues: joint value creation in a network and the fact that consuming is also considered a value-adding activity, indicating that the consumer might be part of the network and is involved at more than just the output side.

4.2.2.16 Müller-Stewens and Lechner (2001)

The authors approach the definition of a business model by extending the "value creation model" which depicts the central activities of a business and their systematic interplay (Mller-Stewens and Lechner 2001, p. 374). This model has to be considered and concretized from a capitalization perspective. This capitalization results from four partial models (Mller-Stewens and Lechner 2001, p. 410):

- The value offering model allocates the product/service portfolio of a company to the targeted customer segment. Furthermore, the needs of these segments are considered
- The value creation model shows the structure of the value creation (resources, make-or-buy-decisions, tasks of partners, internal communication channels and

coordination mechanisms). Here, the IT infrastructure is considered a critical success factor
- The marketing model defines the relation between customer and company
- Revenue model

According to authors, the business plan is not an integral part of the capitalization, though "completing" the business model. Hence, the authors blur the boundaries between these two concepts which is uncommon in business model or business plan literature respectively.

4.2.2.17 Weill and Vitale (2001)

Weill and Vitale (2001) talk about different abstraction levels when introducing the concept of e-business models. Atomic e-business models describe the very essence of how business is conducted. Each of them is characterized by four elements: "strategic objectives, sources of revenue, critical success factors, and core competencies" (Weill and Vitale 2001, p. 25). The actual e-business model combines atomic e-business models, describing a firm's business activities in a particular initiative (i.e., in a particular business segment). This e-business model enriches the atomic e-business models by adding roles and relationships among a firm's customers, allies, and suppliers. Furthermore, product, information, and financial flows are being summarized, the latter related to the participant's benefits. Closely related to the e-business model, but as itself not part of it, is the e-business initiative which is the combination of the above-mentioned e-business model and the target customer segment, distribution channels, and the IT infrastructure. Finally, the e-business implementation completes the e-business model and the e-business initiative by considering factors like financing, recruitment, marketing, and incentives. Weill and Vitale's perspective is summarized in Fig. 4.3.

4.2.2.18 Wirtz (2001)

According to Wirtz (2001, p. 81), the business model comprises of the illustration of the workflow as well as the flow of materials and information both within the organization and between the organization and external stakeholders, and finally the stream of revenues. Business models constitute an aggregated presentation and conceptualization form, consisting of six partial models that combine to an integrated business model.

- The market model can be divided into competition and buyer models and describes the mutual forces of the organizational environment as a market and the individual buyer.
- The procurement model describes how to organize the company's input factors. Input factors are in particular production factors that are influenced by the previously mentioned market behavior and market structure.
- The value creation model describes the combination of goods and services and their transformation into offerings with emphasize on input/output ratio.

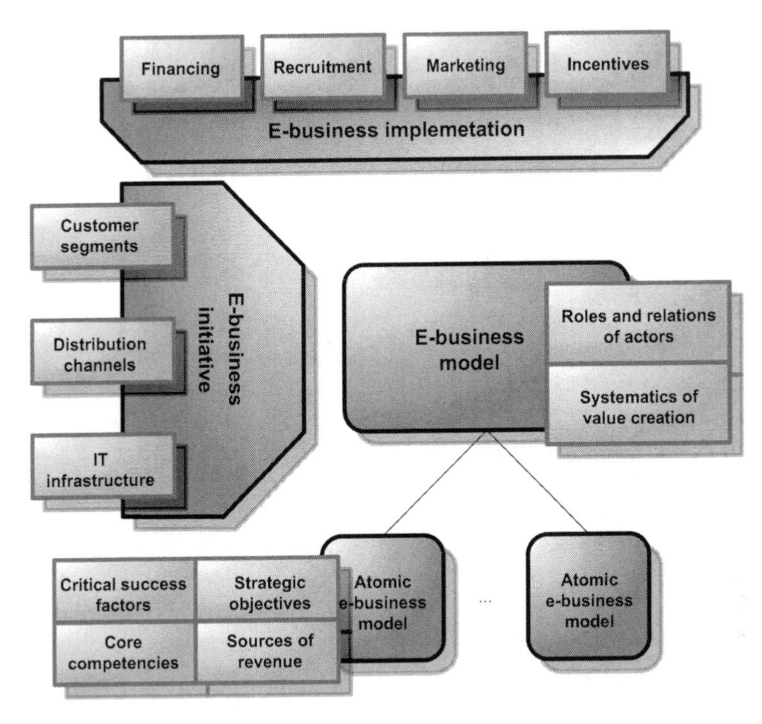

Fig. 4.3 E-business model according to Weill and Vitale (2001). Adapted from Weill and Vitale (2001)

- The value offering model shows which goods and services are being offered to which customer segment.
- The distribution model links to the previous model by asking the subsequent question of how and under what conditions the respective goods and services are made available to the customer. It is important to make a distinction here between tangible products that need to be distributed physically on the one hand and intangible products or services that can be transmitted electronically on the other hand.
- The capital model can be regarded as the most important of the six sub-models. It illuminates financing opportunities and feasible revenue generation models. Special attention should be paid to the various revenue generation models. These are the core of an enterprise in electronic business (Wirtz 2001, p. 8).

4.2.2.19 Yu (2001)

"The [e-commerce] [...] business model can be defined as a conceptual architecture for representing entities and relationships of model components with identified critical success factors of electronic businesses" (Yu 2001, p. 114). Importantly,

Yu (2001) limits the business model to components that are crucial for the success of a company. He identifies fifteen key components of business models: Market and customer segmentation, role of customer, product description, service description, assets (including raised money, technology infrastructure, patents, knowledge, expertise, and the likes), costs, pricing strategies, promotion as specific marketing activity, distribution channels, revenues and its sources, profits, market share, economic scale, marketing strategies and plans, and competitive advantage.

Obviously, Yu (2001) depicts a very detailed view on business models, integrating aspects. Furthermore, the definition includes a multitude of related concepts, which are oftentimes considered not to be part of a business model, such as business plan or marketing model. Interestingly, the author's view on the business model definition changes in the course of the article: contrary to his statements in the initial part of the article, marketing strategies and plans as well as competitive advantages are explicitly considered to be elements excluded from the business model itself (Yu 2001, p. 118).

4.2.2.20 Bieger et al. (2002)

Bieger et al. (2002) identify eight trends of the modern net economy each of which is mapped into a business model element. They stress that the focus cannot be on single companies any more, but on networked organizations (or even whole industries): A business model is the way a company, a networked organization, or an industry creates value, giving answers to the following partial models (Bieger et al. 2002, p. 50):

- *Value offering concept:* Definition of target customer segment and value proposition
- *Communication concept:* Retention of the goods or services in the customer segments to establish long-term customer relations
- Revenue concept for the whole network
- Growth concept
- *Configuration of competencies:* Core competencies and their interaction in the network as well as identifying areas where external know-how has to be purchased. Furthermore, the mechanism for partner management has to be settled
- *Organizational form:* Positioning in the value chain according to above-mentioned core competencies
- *Cooperation concept:* Choice of cooperation partners. It does not become clear whether the authors refer to cooperation partners within the network (from each company's point of view) or network-external partners
- *Coordination concept:* Choice of coordination form

The extensive definition given by Bieger et al. (2002) are mainly from a whole network perspective, emphasizing the importance of joint value creation and the associated challenges in terms of coordination and communication. Herein, some of the proposed elements are very basic issues of s network's overall strategy.

4.2.2.21 Chesbrough and Rosenbloom (2002)

Chesbrough and Rosenbloom (2002) consider the business model as a mediating construct between technology and economic value. They put forth six functions of a business model[5]:

- The value proposition is the value which is created for users by the product or service offering.
- The market segment highlights the user group that is targeted by the offering.
- The structure of a firm in the value chain required to create and distribute the offering.
- The cost structure and profit potential of producing the offering.
- The position of the firm within the value network links suppliers and customers, including identification of potential complementors and competitors.
- The competitive strategy ensures that the firm will gain and hold advantage over rivals.

4.2.2.22 Magretta (2002)

Business models are "stories that explain how enterprises work" (Magretta 2002). They accentuate potential customers and their valuation as well as the revenue model of a business. Furthermore, the business model outlines the economic logic that shows how to deliver value to customers at an appropriate cost, that is the way value is created efficiently. Magretta sees every business model as a variation on the "generic business value chain underlying all businesses" (Magretta 2002). This general logic is divided into (1) producing and (2) selling products or services.

Importantly a "business model isn't the same thing as a strategy, even though many people use the terms interchangeably" (Magretta 2002). Business models rather ought to show the interplay of the elements of the whole system, implying at least the consideration product and information flows as well as the various actors and their coordination. Processes of value creation might also be a part of this interplay. Hence, business models do not factor competition, which is in fact a crucial performance factor – it is in fact a matter of strategy to deal with these kinds of reality. "Doing better than your rivals" (Magretta 2002), that is being able to differentiate oneself from competitors, is a matter of competitive strategy.

4.2.2.23 Osterwalder and Pigneur (2002)

Along the lines of Petrovic et al. (2001), Osterwalder and Pigneur (2002) position the business model between strategy and implementation into processes. The business model describes "the value a company offers to one or several segments of

[5]The definition was initially set up as a working paper (Chesbrough and Rosenbloom 2000).

customers and the architecture of the firm and its network of partners for creating, marketing and delivering this value and relationship capital, in order to generate profitable and sustainable revenue streams" (Osterwalder and Pigneur 2002). The e-business ontology developed in Osterwalder and Pigneur (2002) is founded on four pillars:

- Products and services offered by the firm, representing the value proposition, the target customers, and the capabilities of the firm itself.
- The infrastructure and the network of partners describes the activity configuration, the resources and assets, and a firm's partner network necessary to fulfill the value proposition.
- Customer relationship defines the distribution channels and the strategy to retain customers' trust and loyalty.
- The financial aspect is composed of the revenue model and the cost structure of a firm which result in the profit model.

Even though the authors introduce a definition for firms, they highlight the importance of a partner network. Another important notion is the high level character of a business model, stating that detailed process descriptions cannot be given in a business model.

4.2.2.24 Stähler (2002)

Based on a brief literature overview, Stähler (2002, p. 41) develops a tripartite definition comprising the value proposition, the configuration of value creation, and the revenue model.

- The value proposition describes the utility, that is the value for customers as well as partners which is created by applying the business model. The value proposition for customers describes the benefit gained by the customer. Thus, it is a determination of customer needs to be fulfilled, implicitly showing the focusing of the company. The value proposition for partners describes the benefits gained by partners contributing to the value creation process, revealing their intrinsic motivation to participate and become a part of the business model.
- The architecture of value creation is composed of the product and brand design, the internal architecture, and the external architecture. The product design is the configuration of the product which is offered to fulfill its value proposition. By offering this very product, the company can set apart from its competitors. Consequently, an explicit description of the competitive landscape is not part of a business model. Rather, the business model answers the question of how a company can differentiate itself from others. Furthermore, targeted market and customer segments are described. The internal architecture of value creation is composed of the input resources (core competencies and assets), the transformation process and the coordination mechanisms. The external

architecture comprises the customer interface (communication and distribution) and a description of external partners. Apparently, there are overlaps between the internal value creation process and the role of external partners.

- The revenue model reveals the sources of revenue and the way a company earns profits (that is, which are the pricing mechanisms and who pays?).

4.2.2.25 Pateli and Giaglis (2003)

Pateli and Giaglis (2003) design a generic framework which is based on a state-of-the-art analysis of literature. They provide a two-dimensional framework extending the model of (Alt and Zimmermann 2001). The horizontal frame includes the "primary components" of a business model (Pateli and Giaglis 2003, p. 338):

- Mission (strategic objectives)
- Target Market
- Value Proposition
- Resources (capabilities and assets)
- Intra- and inter-organizational processes
- Cost and revenue models including pricing strategies
- Value chain/net (partnerships)

The vertical frame is comprised of what the authors call underlying components, which consist of enablers, influencing and limiting factors: Market trends, regulation, and technology (cp. Fig. 4.4).

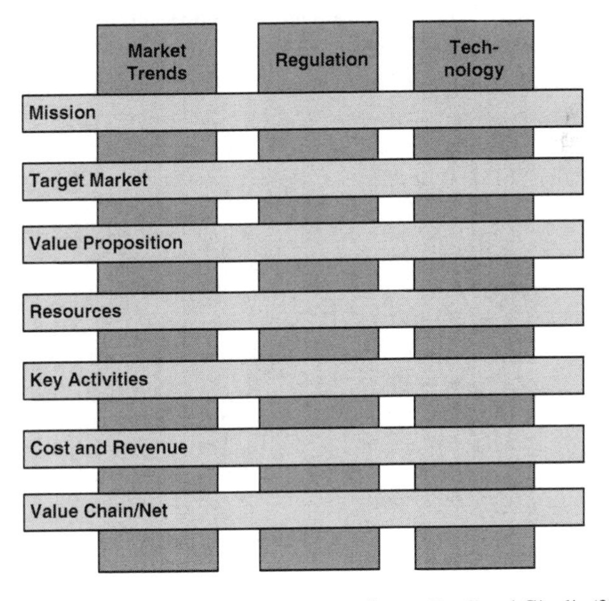

Fig. 4.4 Business model components framework according to Pateli and Giaglis (2003). Source: Adapted from Pateli and Giaglis (2003)

4.2.2.26 Scheer et al. (2003)

Scheer et al. (2003) distinguish between business models and Internet-based business models after comprehensively analyzing the state of the art literature. Business models describe in a simplified way ("bird's eye view") the regular business conducted by an organizational unit (i.e., an independent profit center). This business consists of involved actors (or roles), transformation processes reflecting the value creation, transfer streams (product, information, and financial flows), influencing factors from the market (e.g., competitive forces, changes in the customer segments), and auxiliary means (Scheer et al. 2003, p. 22). The latter is the only explicit differentiation criteria of business models and internet-based business models, namely the internet technology as enabler. Contrary, technology can also be used in traditional business models (Scheer et al. 2003, p. 21) such that a differentiation generally seems to be not necessary. Despite conducting a comprehensive literature analysis, the authors do not consider the revenue model, just name financial flows as compensation for delivered goods or services. On the other hand, factors like influencing factors from the market and auxiliary means are being interpreted as integral parts of a business model though oftentimes being considered external forces imposing pressure upon the business model from outside (as, for instance, postulated by Osterwalder (2004)).

4.2.2.27 Osterwalder (2004), Osterwalder et al. (2005)

Osterwalder's contribution to business models research is twofold: On the one hand, it clarifies the concept of business models by outlining the differences to related concepts, on the other hand an ontology to "conceptually express the business logic of a firm in a structured form" (Osterwalder 2004, p. 5) is created. Due to the latter purpose, Osterwalder (2004, p. 42) erects a very detailed framework to capture the business model of a firm as a basis for his ontological implementation. The author identifies four areas that should be addressed by a business model:

- *Product:* A business model has to explain the type of business a firm is in, the very goods and services being offered, and the value proposition.
- The Customer Interface identifies the target customers, the distribution logic and channels, and the customer relationship management.
- The logic of the value creation in a network of partners, including infrastructural and logistical issues is being outlined in the Infrastructure Management.
- *Financial Aspects:* The business model's sustainability from a financial point of view is described by its revenue model and cost structure.

In order to evaluate these four "pillar" more detailed, nine partial models (building blocks) of business models are being introduced (cp. Table 4.1).

With respect to the clarification of the concept of business models, Osterwalder identifies elements that are important business aspects related to the competitive

Table 4.1 Building blocks of a business model according to Osterwalder (2004)

Pillar	Building block
Product	• Value proposition: Overall view of the portfolio offered by the firm
Customer interface	• Target customer: Customer segment the created value is offered to • Distribution channel • Relationship: Link established between customer and firm
Infrastructure management	• Value configuration: Arrangement of activities and resources necessary to create value • Capability • Partnership: Co-operations of the firm initiated in order to create value
Financial aspects	• Cost structure • Revenue model

landscape and the implementation of a business model, though itself not being a part of the business model:

- Capital model as external source of capital
- Market model as means to situate the business in the competitive landscape, including strategic considerations like the company's positioning with respect to rivals, potential new entrants, or substitutes
- ICT is considered an enabler for business models, but not an integral part of it

4.2.3 *Quantitative Analysis of the State-of-the-Art Business Models*

In order to grasp the contributions of the state-of-the-art literature presented in Sect. 4.2.2, a matrix allocating named aspects to categories is set up. The rows show the contributions to scientific literature. As indicated in Sect. 4.2.1, contributions are incorporated that provide either a definition or a decomposition of the concept business models into partial models. The columns indicate specific components of the business model definition which were assigned to six categories value creation, network, customers, profit, market view, and miscellaneous. The categories and their characteristics are depicted in Table 4.2.[6]

Via the matrix, the different contributions to business model research are allocated to addressed characteristics. In order to visualize the statements of the authors, four different colors are utilized. Dark-green coloring indicates the explicit definition of building blocks whereas light-green is used for implicitly named characteristics. Furthermore, dark-red coloring denotes that this component is

[6]Note that the final building blocks for this book's business model definition will be introduced in Sect. 4.4.2.2. The components introduced above are only for the purpose of analyzing the state-of-the-art literature.

Table 4.2 Categories for the qualitative analysis

Interim category	Characteristics	Description
Value creation	Value creation and offering	Description of products or services offered; the value is created by transforming a given initial state into a (preliminary) final state (Scheer et al. 2003, p. 21).
	Value creation process	Not only the description mentioned above, but also a more detailed overview on the underlying processes. Closely related to product/service flows.
	Value proposition	The value proposition is, similar to the value creation, but much more condensed and translated into the customer view, an overview on the portfolio of goods and services offered (and that create value for the customer) (Osterwalder 2004, p. 44).
	Product/service flows	Dependencies in the value creation process, showing the arrangement of activities and resources needed to create value.
	Information flows	Exchange of information that is required within the value creation process.
	Success factors of product	Explanation why a product or service is able to differentiate against competitors offers and to ensure a sustainable success.
	Core competencies	Major capabilities of the organizational unit considered.
Network	Business partners, actors, and their roles	Depending on the scope of consideration, the actors can be either partners from a single companys perspective or the participating actors that constitute a value network.
	Position in value system	Closely related to the actors and roles. The position in the value system indicates what the company itself contributes to the goods and services offered by the value network. Thus, theres also a connection to the core competencies.
	Coordination of actors and activities	Also closely connected to product/service flows, additionally incorporating the actors.
	Know-how of employees/assets	Assets can be financial, intangible such as patents or copyrights, but also human, outlining the knowledge of the employees (Afuah and Tucci 2003, p. 69).
	Incentives	Motivation for actors to participate in the value creation process.
Customers	Communication with customers	Description of link established to the customers.
	Customer-centricity	Highlighted position of customer as center of the value creation. The customer is not only situated at the output side, but also involved in the value creation process (cp. Sect. 4.3).
	Customer segments	Addressed target group.
	Distribution (channels)	The channel via a customer is approached.

(continued)

Table 4.2 (continued)

Interim category	Characteristics	Description
Profit	Revenue	Description of the way revenue is generated.
	Benefits for actors	Closely connected with value proposition, benefit/utility provided by the goods and services offered, for internal and external actors.
	Financial flows	Transfer flows of the actors to requite obtained goods or services (both internally and externally).
	Costs	Consideration of the cost structure of the business, e.g., caused by transactions or coordination.
	Funding (external sources)	Sources of funding from external investors used to finance the business activity.
Market view	Market consideration	Description of the environment of a firm, e.g., suppliers, competitors, complementors, possible entrants, etc.
	Marketing	Description of marketing strategy of the organizational unit under consideration (Timmers 1999, p. 32).
	Competition	Sub-characteristic of market consideration, merely relating to the competitive enviroment.
Misc	Technologie	Technology as enabler for value creation.
	Strategic objectives	Description of the mission, i.e., the strategic alignment of the organizational unit.
	Business model = strategy	Equalization of "business model" and strategy or the statement that strategy is a part of the business model.

explicitly excluded from the business model definition, analogously a light-red flag denotes an implicitly mentioned negative nomination. Explicit mentions are counted fully, implicit ones are counted as 0.5. In addition, the matrix is enhanced with some additional information on the source analyzed:

- A specification of the author's objective, namely the provisioning of a general definition or a decomposition of the concept business model into partial models (1), devising a taxonomy that outlines generic types of business models (2), or the investigation of specific real-world business models (3). Contributions to literature[7] providing solely formalisms for representing business models, change and adaptation methodologies, or evaluation models without introducing new aspects (Pateli and Giaglis 2003) were not considered.
- The research background of the author, displaying the discipline which motivates the objective of the respective contribution. Contributions from technically motivated authors (T), authors situated at the interface between business and internet topics (eB), and authors mainly concerned with business strategy (S) could be identified.

[7] Cp. e.g., Papakiriakopoulos et al. (2001), Schwickert (2004).

- Source of the publication: Book (B), dissertation (D), book chapter (C), journal (J), conference or workshop proceedings (P), working paper (W), or merely electronically published (E).

The literature review reinforces the statement of several authors saying that the general understanding of the term "business mode" is still rather discordant about what constitutes a business model. The portfolio of definitions ranges from more tightly drawn concepts (cp. e.g., Magretta (2002); Osterwalder (2004); Timmers (1999); Weill and Vitale (2001)) explicitly factoring out market consideration to very comprehensive perceptions (cf. e.g., Afuah and Tucci (2003); Hamel (2002); Petrovic et al. (2001); Wirtz (2001); Yu (2001)) which, partially, even include the business plan as an integral part of the business model.

On the other hand, the pool of scientific contributions provides both high-level definitions with low granularity (cp. e.g., Gordijn and Akkermans (2001); Lindström (1999); Nilsson et al. (1999) and highly detailed disquisitions on the components of a business model (Afuah and Tucci 2003; Hamel 2002; Osterwalder 2004). The detailed allocation of business model components to the characteristics introduced in the previous section is visualized in Fig. 4.5.[8]

Yet the range of considered literature is broad, a certain subset of common elements can be identified. Based on the number of nominations in literature, it can be constituted that the categories value creation, network, customer, and profit are generally well-established components of business models.

- *Value creation:* Doubtlessly, some description of what is actually offered to the customer is an essential part of a business model. Dissent can be observed in the question of how detailed such a description should be. While some authors leave it at a more marketing-biased value proposition, a large share of authors call for the description of dependencies and activities performed in order to create value. Twelve contributions see a description of the processes underlying the value creation as an integral part of business model. However, the requested granularity differs significantly, from a high level overview to a detailed description of business processes.
- *Network:* The description of business partners, actors, and their roles is supported by the majority of authors. The bulk of the literature speaks of business models for individual companies. However, actors are sometimes also interpreted as universal organizational units. Depending on the perspective, business models can apply for both single companies and value networks (Bach et al. 2003; Buchholz and Bach 2001; Scheer et al. 2003).
- *Customer (and market):* As the value creation process is finally adjusted to fit customer needs, a certain integration of the link established to the customers, addressed segments, or distribution channels are oftentimes mentioned (cp. e.g., Afuah and Tucci (2003); Bieger et al. (2002); Chesbrough and Rosenbloom

[8]Bartelt and Lamersdorf (2000) and Bartelt et al. (2001) are not explicitly listed in the matrix overview due to their analogy to Timmers (1999).

Group	Component	Sum positive	Sum negative
Value creation	Value creation and offering	22,5	0
	Value creation processes	12	1
	Value proposition	14,5	0
	Product flows	15	0
	Information flows	9	0
	Success factors of product	3,5	0
	Core competencies	7,5	0
Network	Business partners, actors & their roles	21	0
	Position in value system	14,5	1
	Coordination of actors and activities	14,5	0
	Know-how of employees/assets	8	1
	Special emphasize on partnerships	9,5	0
Customers	Communication with customers	11,5	0
	Customer-centricity	1,5	0
	Customer segments	13	1
	Distribution (channels)	11	1
Profit	Revenue	20	0
	Benefits for actors	6,5	0
	Financial flows	12,5	0
	Costs	8,5	0
	Funding (external sources)	2	2,5
Market	Market consideration	7,5	3
	Marketing	4,5	2,5
	Competition	5	3,5
Misc	Technology	3	6
	Strategic objectives	7,5	2
	Business model = strategy	1	5

	Purpose of contribution	Background	Source
Timmers (1998)	12	eB	J,B
Lindström (1999)	(1)	T	C
Nilsson et al. (1999)	(1)	T	C
Heinrich (2000)	1	eB	C
Kluber (2000)	1	eB	P
Mahadevan (2000)	12	eB	J
Zimmermann (2000)	1	eB	P
Afuah, Tucci (2002)	1(2)	eB	B
Alt, Zimmermann (2001)	1	eB	J
Amit, Zott (2001)	1	eB	J
Buchholz, Bach (2001)	1	S	W,C
Gordijn, Akkermans (2001)	1	T	P
Müller-Stewens, Lechner (2001)	1	S	B
Rappa (2001)	2	eB	E
Weill, Vitale (2001)	12	eB	B
Wirtz (2001)	123	S	B
Yu (2001)	1	eB	P
Bieger et al. (2002)	1	S	C
Chesbrough, Rosenbloom (2002)	13	eB	J
Hamel (2002)	1	S	B
Magretta (2002)	13	S	J
Osterwalder, Pigneur (2002)	1	T	P
Stähler (2002)	1	eB	B
Pateli, Giaglis (2003)	1	eB	P
Scheer et al. (2003)	1	eB	W
Osterwalder (2004)	1(2)	T	D,J

Caption:
- explicitly named component of a business model
- implicitly named component of a business model
- component explicitly excluded from the business model definition
- component implicitly excluded from the business model definition
- ambivalent statements

Fig. 4.5 Quantitative state-of-the-art analysis of business model literature. Source: Design adapted from Scheer et al. (2003)

(2002); Hamel (2002); Heinrich and Leist (2000); Klueber (2000); Osterwalder (2004); Stähler (2002)). In this connection, only a few authors strictly interpret customers as part of the market(ing) model and therefore strictly differentiate a business model from measurements to reach customers which include for example distribution channels or other links (cp. Timmers (1999); Weill and Vitale (2001)). Others deny an extensive consideration of the market including competitors, possible entrants, etc., but concede as true that certain relations to the customer do not only impact a business model, but are an integral part of it (e.g., the determination of target segments, distribution channels, and integration of customers in the value creation process) (Bach et al. 2003; Magretta 2002; Osterwalder et al. 2005; Osterwalder 2004).

- *Profit:* The business model is oftentimes referred to as a description of how a company earns money (cp. e.g., Afuah and Tucci (2003); Rappa (2001)). Several publications stress the revenue generation as one of the most important components of a business model (cp. e.g., Stähler (2002); Wirtz (2001)). Facet of financial aspects are covered in 22 out of 26 references sources. The cost structure as a counterpart of revenue generation was neglected in earlier literature, but significantly gained importance in later contributions to business model research.

Accordingly, after having gotten a detailed overview, it can be stated that the above-mentioned elements value creation, (partner) network, customer, and profit shall be compulsively included in the business model framework. However, a consistent and accurate business model definition cannot be established without thoroughly scrutinizing what to include into the very business model essence and what to be left aside as a related concept.

4.2.4 Concepts Related to But Not Part of a Business Model

A striking insight provided by the literature review is not only the disunity among authors concerning related elements, but also the blurry notions about what actually belongs to a business model, i.e., elements prior to or subsequent to the business model (e.g., technology, marketing, strategy, business plan, funding, or legal aspects).

4.2.4.1 Differentiation from Business Strategy

The differentiation between business models and the business strategy is extensively discussed in literature. According to Stähler (2002, p. 48), a business model is not a strategy. Magretta agrees, stating that "a business model isn't the same thing as a strategy" (Magretta 2002, p. 89). Business models do not consider competition; determining how to perform better by differentiating from competitors is addressed by the strategy. Consequently, the strategy can be considered as a precondition for business models, outlining the vision and objectives of a company (Osterwalder 2004, p. 14). Most commonly, strategy is defined as a design for action, a deliberate plan before starting a concrete course of action (Mintzberg 1987, p. 11). Thus, the business strategy is situated one level above the business model.

4.2.4.2 Differentiation from Implementation

Alike, the strategy of a firm is the basis for a business model on the one side, its implementation should also be excluded from the business model concept. The organizations and workflows that are necessary to execute a business model should be located to the business process layer, not to the business model layer. Osterwalder et al. (2005) put it as follows: "[A] business mode cannot be successful per se" although a couple of authors speak of successful business models. Oftentimes, the business model as itself is not enough to illustrate a company's success. When it comes to sustainable market success, the market(ing) model can be the differentiating element (Sandrock and Weinhardt 2006; Timmers 1998). Furthermore, part of the implementation is the relationship management among partners based on trust (Allee 2008; Bieger and Regg-Strm 2002).

4.2.4.3 Differentiation from Business Plan and Business Case

In case of considering a start-up, after elaborating a business model, a business plan is set up. A business plan finally justifies to external sources, e.g., investors[9] or analysts from rating agencies, that a business model is elaborate, feasible, and finally profitable (Stähler 2002, p. 48). Hence, a business plan can be interpreted as an intermediate step between the conceptual business model and its implementation, consulting competitive considerations made in a company's strategy. The content addressed in a business plan exceeds the scope of a business model in several areas. The business model itself does not explain how an organizational unit implements its vision or objective, respectively (Timmers 1998, p. 32). This is concretized in the marketing strategy which is amongst other things (e.g., funding) outlined in the business plan. Thus, a business plan aims to draw a broad picture of the economic perspective and the value creation of an organizational unit (Brugger 2005, p.30).

Several criteria of the business plan do not directly address the business and the participating actors themselves, but are focused on the implementation of a business (model) by considering external factors which determine the success (or failure) of a business model. For instance, the business plan introduces the marketing strategy in a much more detailed and holistic level than the business model, which only singles out specific factors that are directly connected to delivering value to customers. Moreover, the business plan pinpoints a realization schedule, risk analyses, and sources of funding, which are in turn not directly allocated to the business model, but rather consider its implementation and environment.

The business case is another instrument which is to be differentiated from a business model or a business plan. A business case is considered a scenario to economically evaluate an investment (Brugger 2005, p. 11). Hence, the scope of a business case is much more detailed than the scope of a business model. It highlights

[9] Venture capitalists or outside creditors as well as investors when it comes to going public.

a single project, whereas business model and business plan picture a generalized view on the whole organizational unit.

4.2.4.4 Differentiation from External Forces

Not only the business model, but also strategy and implementation are subject to external forces. This environment is comprised by three kinds of influencing factors: Regulation and legal environment, market trends, and technology (Alt and Zimmermann 2001; Osterwalder 2004; Pateli and Giaglis 2003; Papakiriakopoulos et al. 2001). Through changes in legal environment, business models can, for instance, turn illegal. Especially, changes in Internet-related regulation, e.g., privacy, must be continually considered by companies in the e-business sector. The market forces emanate from two sources: customer and competitors. The competitive environment as an external factor and the business model are strongly linked since it is the business model which enables a firm to have competitive advantage. Technological change enables companies to exploit whole new business opportunities, not only in the value creation process, but also by providing new billing opportunities, distribution channels, or ways to communicate with partners. However, especially the link between business models and ICT is obviously very strong. With the emergence of e-business models, the business model cannot be decoupled from technology anymore. ICT can be both an essential part of the value creation (e.g., an online auction) and an enabler for innovative business models, e.g., facilitating the communication in large and globally distributed partner networks. So, the environment does not only influence the business model in a competitive manner, but also in a macroeconomic fashion, exerting pressure through government policies, regulation and deregulation, as well as technological improvement.

4.2.4.5 Summary

Following the differentiations made in this section, an integration of the business model in an overall view on a firm or a network, respectively, is possible, and thus clarifies the business model's role with respect to related concepts. These dependencies are illustrated in Fig. 4.6 via a "business stack." Roughly spoken, the business model closes the gap between more abstract strategic issues and detailed business process considerations.

4.2.5 Typologies of Business Models

Typologies of business models are not relevant for the remainder of this book and are, therefore, not discussed in more detail. However, in order to round off the integrated view on the topic "business model" as illustrated in Fig. 4.1, business model typologies are briefly sketched in the following.

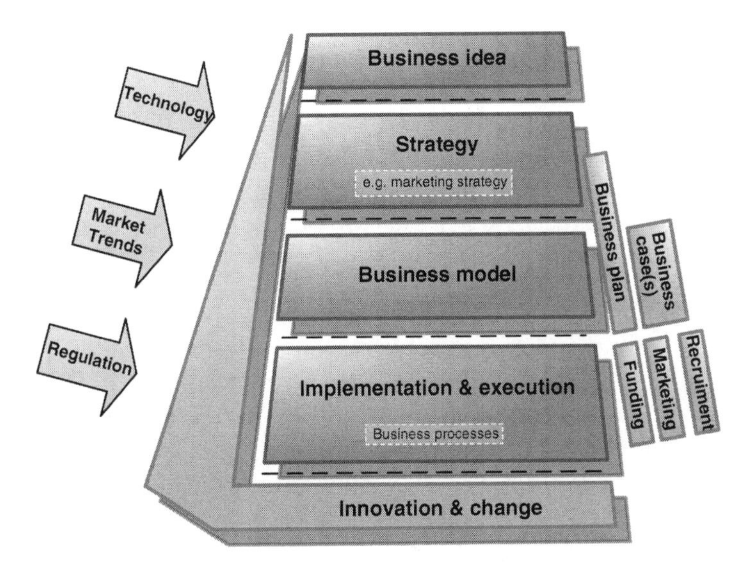

Fig. 4.6 The business stack

Business model research is not only about defining the concept and decoupling it into partial models or components, respectively. A considerable number of authors[10] also dealt with classifying business models, thus providing generic business models that are mostly categorized on the basis of specific criteria (Pateli and Giaglis 2003). Business models are abstract, rather complex concepts which can be concretized by developing a classification scheme (Lambert 2006).

The criteria used for the classification of business models differ, oftentimes they seem to be more intuitively chosen than based on a specific business model component. The question which criterion to use for a distinction of business model typologies cannot be answered in general. Wirtz (2001) suggests the usage of the value created as classification criterion since it provides homogeneous classes by capturing a lot of "pure players" offering an exactly circumscribable value, whereas other criteria oftentimes lead to duplications. Moreover, the concrete business models assigned to the classes would be more heterogeneous than in the case of using the value created as classification criterion.

4.3 Customer Integration

While customer-centricity as a pervasive occurrence of customization has already found its way into literature, real customer integration, that is, any form of integration into value creation processes that exceeds the "mere" customization, e.g.,

[10]Cp. e.g., (Afuah and Tucci 2003; Eisenmann et al. 2001; Linder 2000; Rappa 2001; Timmers 1998; Weill and Vitale 2001; Wirtz 2001).

through configurators or requests to mediators, is frequently postulated (Tapscott and Williams 2006), yet still in its infancy. Advances in ICT already altered and still have a huge potential to radically change collaboration patterns. Collaboration approaches in heterarchies and market structures supersede firms being organized in fixed hierarchies. New business and organizational models, respectively, tend to put more emphasis on community, collaboration, and self-organization, at the same time disregarding hierarchy and control. Customer integration is thus the main manifestation of newly appearing business networks and is essential for the automated provisioning of complex services.

Let us consider the fundamental idea of outsourcing, at heart going back to Coase (1937). Very simply spoken, as long as the administrative and coordinative costs for another transaction within an organization is cheaper (or "better") than a transaction at the open market, a firm will not outsource any processes. As soon as this threshold is reached, outsourcing is profitable. As mentioned earlier, collaboration among firms is already established, evolving from fixed contracts to loosely coupling (cp. Chap. 3). So, what can already be seen happening in inter-organizational collaboration is still uncommon in the business-to-consumer relationship.

4.3.1 Two Facets of Customer Integration

Fundamentally, customer integration has two facets. On the one hand, the term *prosumption* is used when customers are directly connected with the product or service offered. On the other hand, *crowdsourcing* approaches consult private individuals that are not directly connected to the product or service as "external workforce."

4.3.1.1 Prosumption

This type of involvement can be considered the silver bullet of customer integration since people with a directly connected, oftentimes emotional binding to the offered goods or services. To describe this mode of collaboration, the term "prosumption" is oftentimes consulted. Interestingly, it was in the pre-Internet era, prior to the possibility to simply access a huge pool of possible ideas, innovations, development potential, and the likes via the World Wide Web, when Toffler (1980) coined the illustrative term prosumption, a made-up combination of the terms production and consumption. Today, the term describes a situation in which customers co-create and co-innovate goods and services rather than simply consuming the end product (Tapscott and Williams 2006, p. 1). Moreover, prosumption goes beyond mass customization or customer-centricity with the tailoring of a product or service by composing inherently standardized modules. Prosumption is about consumers having a "genuine role in designing the products of the future [...] in their own networks, and for their own ends" (Tapscott and Williams 2006, p. 149). Especially, the last aspect is important: customers participate for their own ends, meaning that an involvement is associated with personal advantages.

As an example, let us consider Lego's Mindstroms,[11] Lego toys with programmable sensors, motors, and controllers. To deeply involve the community – which turned out to be built up by lots of adults, not only by the primarily targeted group of teenagers – Lego offers a free software development kit to modify the programmable parts in the toys.[12] As a result, users post their creations on the website to other community members who can order this bundle made up of bricks, parts, and instructions. Each new customer-generated application enhances LEGO's product portfolio, making the toy more and more valuable. When Lego planned to develop the next generation of the toy, Mindstorms NXT, they hired four of the most prolific users as de-facto employees for the development cycle (Tapscott and Williams 2006, pp. 130–131). Moreover, the Mindstorms Developer Program is a sub-community of 100 pioneers from around the globe who design kits that are not only purchasable via the website, but also are officially released. The advantages for LEGO are obvious: On the one hand, listening to customers cannot be take place more directly – which lets LEGO react quickly on demands and, thereby, increases customer satisfaction. On the other hand, the concept saves resources since LEGO (at least partly) outsources the job of product managers to its community.

4.3.1.2 Crowdsourcing

The second facet of customer integration is usually circumscribed as open innovation (Chesbrough 2003; Chesbrough et al. 2006) or the made-up term crowdsourcing, which is assembled from "crowd" and "outsourcing." These concepts denote the same idea as prosumption, but addressing a different scope or rather, target audience. Corporate tasks and structures are not outsourced to other companies but to an anonymous crowd of spare time laborer. The objective is to inspire an undefined, large community of customers and users to generate solutions to problems previously processed internally or by professional outsourcing providers. Mostly, product management tasks and research tasks as well as unsolved problems are being crowdsourced. That is, tasks are outsourced to private individuals that are not directly bound to the product or service, but rather interested in the domain at hand in general.

While LEGO's integration model is based on non-monetary incentives – e.g., own benefits and reputation in the community, crowdsourcing approaches are oftentimes compensated monetarily. For example, InnoCentive[13] is an open innovation marketplace providing a platform for companies (so-called "seekers") to outsource unsolved challenges to a solver-community who can earn a considerable amount of money for a solution. According to statistics, InnoCentive unifies 225,000 solvers

[11] http://mindstorms.lego.com/.

[12] The Mindstorms software license includes a "right to hack."

[13] http://innocentive.com/.

dedicated to different disciplines.[14] Renowned companies like Procter & Gamble are registered as seekers, profiting from the pool of knowledge, "collective brainpower," that is located outside of the company boundaries.

The above examples illustrate that ICT facilitate not only the appearance new products and services, but that they require the adjustment and changes to the entire business models.

4.3.2 *Interim Conclusion*

Prosumption as well as crowdsourcing particularly suggest themselves in business networks for two major reasons:

- Usually, a large customer base is addressed by the value proposition of such networks, which is much larger than any customer base that could be addressed by a single company. Therefore, participants in business value networks are not only able to leverage the capabilities of partners, they are also able to leverage the knowledge provided by customers.
- Business models of companies operating in business networks are already – by definition – more broad-minded to openness, peering, (knowledge) sharing, or globality, consequently have already overcome resentments in this regard. "Not-invented-here" resistance against external knowledge should be lower due to already opened interfaces to a multitude of partners which may be coupled loosely, not via contracts (Tapscott et al. 2000). Furthermore, the ability to transfer external inputs into internal processes is required in open business networks anyway, integration platforms and the likes may already be in place.

With market trends being one of the major drivers and demanders for business model change and/or innovation, market sensitivity matters – that is an early identification of changing or newly arising customer needs. So, today's companies will have to ask themselves whether they can afford to miss out the opportunity of incorporating consumers into design and innovation processes. In order to capture the potentials of prosumption and crowdsourcing, customer integration is dedicated a separate business coponent (cp. Sect. 4.4.2.1).

4.4 A Business Model Framework for Business Networks

The framework to be presented in this book serves two purposes:

- Clarify the concept "business model" which is still used in several different meanings. It is not the ultimate aim to find the right perspective upon the concept, rather it shall support a common understanding when discussing about business models in a networked environment.

[14] Source: http://www2.innocentive.com/about-innocentive/facts-stats, Status: 31.01.2011.

- Update the state of the art of business model literature which did not evolve as fast as the environment. In a profound analysis of the state-of-the-art business model definitions (cp. Sect. 4.2), it turned out that key elements of business networks are not represented properly in current definitions.

4.4.1 *Shortcomings of State-of-the-Art Business Models*

The extensive literature review has shown that there is in fact a heterogeneous notion of what exactly constitutes a business model, on the other hand, common "pillars" of business models could be identified (cp. Sect. 4.2.3), namely value creation, network, customers, and profit. In detail, several differences have been detected, e.g., strict segregation of market aspects (cp. e.g., Timmers (1998); Weill and Vitale (2001)) versus permitting market-related aspects directly related to conducting the business (cp. e.g., Magretta (2002); Osterwalder (2004)) versus full incorporation of market and marketing issues (cp. e.g., Chesbrough and Rosenbloom (2000); Hamel (2002)). Most of the contributions, however, lack a consideration of interrelations of the business model components.

In Sect. 4.2.4, this book's view on the business model's overall role – based on the literature discussed in Sect. 4.2.2 has already been outlined, segregating it from strategy, marketing measures, implementation, and related concepts, nevertheless pinpointing their interdependencies. This analysis is incorporated in the business model framework.

Eventually, the core characteristics of business networks that include Web services as inherent source of value creation resulting from the analysis performed in Chap. 3 (cp. especially Sect. 3.2 are underrepresented in current business model definitions. The relationship and role of the customer as sketched in Sect. 4.3 is indeed addressed, but still being purely consumer-focused without considering links into the value creation process. The partner network has found its way into several definitions, still oftentimes implicitly assuming that the respective company is the center of the network. In business networks, one has to move away from this notion, clearly understanding that a multitude of partners contributes complementary goods and services to an overall holistic offering. To grasp this situation, own core competencies should be highlighted in the business model while clearly showing the company's position in the value system and the (relevant) components from partners and their roles. This level of partner integration is not included in most of the business model definition.

4.4.2 *Definition of a Business Model Framework for Business Networks*

On the one hand, the framework shall be the basis for elaborating new business models in a business network environment with automatic service composition.

On the other hand, it provides an instrument for the analysis of current networked business models. The business model framework supports the approach of a decomposed definition to ensure a systemized approach when designing or analyzing business models (cp. e.g., Afuah and Tucci (2003); Bieger et al. (2002); Osterwalder (2004); Schwickert (2004); Wirtz (2001)). Moreover, the definition shall not only include components as a loose collection of elements without clearly identifying their interrelation (cp. e.g., Osterwalder (2004); Wirtz (2001); Yu (2001)).

The definition provided in this section follows Osterwalder (2004) and Wirtz (2001) in terms of its structure, providing a clear decomposition in partial models as well as an overview on their dependencies. Five major pillars (partial models) of business models (value configuration model, participant model, offering model, customer model, and profit model) are identified and subsequently refined into 11 business model components. In this context, their dependencies are examined.

4.4.2.1 Pillars of the Business Model Framework

As above-mentioned, five basic pillars of business models for companies are identified which circumscribe the essence of the business performed to a reasonable extent. Only elements that are directly connected with creating, offering, and selling the goods and services provided are considered, abstracting away from related elements that deal with making up a general mission statement (general business idea, strategy), considering the competitive landscape (marketing strategy), and the likes. In particular, a "given environment view" is taken to separate elements directly involved in the provision of goods and services and external factors as well as internal measures that are connected to the success or failure of a business. For instance, the business model does not consider:

- If the business is sufficiently funded when launched
- If the targeted customer group actually is large enough
- If the external visibility is good enough to stand up to providers of substitutes
- If the selling proposition is "unique enough" to prevail in the market

Based on the state-of-the-art review of business model literature and the characteristics of business networks, the following five basic pillars of business models are proposed:

- Value creation model
- Partner model
- Value offering model
- Customer model
- Profit model

The *value creation model* includes the transformation of input factors into product or service offerings. Consequently, this partial model is strongly linked with the *partner model*. It includes sources of procurement (suppliers) as well as actors that are involved in the value creation process including their roles. Additionally, there is also a link into the customer model when it comes to joint distribution of goods and services. The *value offering model* is the primary connector between value creation and providing goods or services to the customer, indicating the business a company is in and the portfolio of goods and services offered as well as the value it provides to customers. Again, a strong link to another partial model is obvious: the *customer model*. At this interface, target customers and the associated distribution channel are determined (i.e., which product or service is offered to which customer on which way?). Additionally, the kind of customer relationship is indicated, i.e., if a company relies on the customer not only as a consumer but also as a product manager or prosumer, there is a link to the partner model as well as to the value creation model. Eventually, the *profit model* highlighting costs as well as revenues is a transversal model influenced by and influencing all other partial models.

4.4.2.2 Components of the Business Model Framework

The above-mentioned business models are refined and decomposed into business model components. Table 4.3 gives a short overview of the proposed components and allocates them to the respective partial model and a detailed description of each component is given. Moreover, the reason for including the component is listed and related elements which were deliberately excluded are named.

Figure 4.7 illustrates the dependencies of business model components.

Table 4.3 Components of the business model framework for business value networks

Partial business models	Business model component	Short description
Value creation model	Value configuration	Arrangement of activities necessary to provide value for the customer.
	Core competencies	Major capacities of an organizational unit.
	Position in value system	Position/role a company takes over in the overall value creation provided by the network.
Partner model	Partners and their roles	Actors involved in the value creation and offering including their roles.
Value offering model	Service/product portfolio	Bundle of goods and services offered to the customer.
	Value proposition	Value proposition to the customer: Which problem of the customer is addressed with the service/product portfolio.
Customer model	Target customer	Group of customers adressed.
	Distribution channel	Channels a product or service is delivered to the customer.
	Customer integration	Role of customer in the value creation process.
Profit model	Revenue	Way of revenue generation and its sources.
	Pricing	Price mechanisms allocated to service/product offerings.
	Cost structure	Expenditures connected with creating, offering, and distributing goods and services.

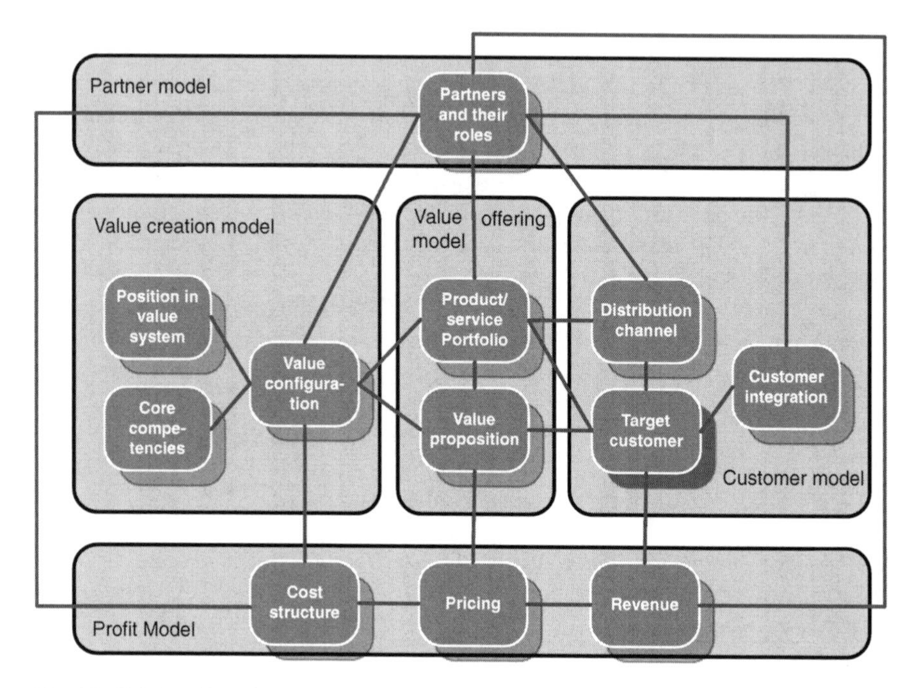

Fig. 4.7 Relationships of business model components

Value Creation Model

The value creation model is called for by several authors (cp. e.g., Buchholz and Bach (2001); Mller-Stewens and Lechner (2001); Osterwalder (2004); Scheer et al. (2003); Stähler (2002); Wirtz (2001)). Three components are included in this partial model circumscribing the process of transforming input factors into a product or service offering that is provided to an end customer or a business customer for further processing.

The *position in the value system* indicates the position a company takes in the overall value creation that is provided by several actors in a business network. Mapping Porter's value chain to the output created by a business network, one identifies the following activities or roles to be performed (Porter 1985, 2001):

- Activities related to the reception, storage, scheduling, and dissemination of input. The collection, analysis, and preparation of information may also be included here ("inbound logistic").
- Any role in the transformation process from inputs into outputs ("operations")
- Activities subsequent to the value creation, such as storage or distribution of output, but also monitoring services ("outbound logistics").

- Marketing and sales activities that provide information for potential customers and auxiliary services inducing customers to purchase offerings and facilitate their purchase, such as information brokering, product/service configurators, or payment services ("Marketing and Sales").
- Any kind of after-sales services to maintain/support the offered product or service ("service").

Besides the above-described dependence upon Porter's value chain, an orientation to a product or service lifecycle may be appropriate in a business network environment.[15]

A company's position in the value system is usually heavily intertwined with its *core competencies*. Competencies or capacities is what follows from resources and assets owned by a company. Resources can be tangible (e.g., equipment), intangible (e.g., patents, copyright, reputation, etc.), or human (skills and knowledge of employees) (Afuah and Tucci 2003, p. 69). The very resources are not listed as individual component, rather being an indicator or cause for capabilities. A competency or a capability can be defined as repeatable pattern of actions that in some way creates value for the customer (Osterwalder 2004, p. 80). On the one hand, capabilities determine a company's place in the value system,[16] on the other hand competencies are a pre-requisite for creating value. The decreasing cost for modularization and subsequent re-configuration to a joint complex product or service offers companies the possibility to concentrate on core competencies and leverage the capabilities of partnering firms in other areas (Hagel III and Singer 2000).

Value configuration is defined following Osterwalder's definition as "arrangement of activities [...] that are necessary to create value for the customer" (Osterwalder 2004, p. 43). It pictures the building blocks that are put together to create the output that is offered in a company's value proposition. Not only the level of added value, but also the sequence and the participating actors should be highlighted, including the most important product/service and information flows. Thus, the value configuration is closely linked with the partner model. A central question is the degree of detail required. As not only the arrangement, but also the sequence of value-creating activity is a frequent call in literature (cp. e.g., Alt and Zimmermann (2001); Amit and Zott (2001); Hamel (2002); Mller-Stewens and Lechner (2001); Stähler (2002)), a high-level consideration of the process can be supported. However, sticking to the notion a business model shall simplify complex real systems, detailed process descriptions cannot be given in a business model (Osterwalder and Pigneur 2002).

---

[15]For instance, a service lifecycle can be composed of the phases *innovation, offering, matchmaking, usage, and feedback* (Stathel et al. 2008), the latter again being closely connected to (re-)innovation (adapted from traditional product value chains, cp. e.g., Kotler and Armstrong (2001)).

[16]Or vice versa, a company entering a new business may acquire its core capabilities according to a desired position in the value system.

Partner Model

Consideration of the partner network is supported by the majority of contributions to literature, yet to a different extent. In this book, the partner model is a separate partial model as it is considered a fundamental part of business models owned by companies in a networked environment. Especially in SVNs, Web services unfold their full potential not before they are offered in a composed manner.

Thus, the partner model can be seen as a transversal model influencing all other partial models. This is not only true in the value creation or offering model. The customer model with the customer becoming a more and more valued partner that is integrated in the value creation process, is also connected to the partner model. Lastly, partners are oftentimes indirect sources of revenue, for instance through advertising or provisions. So, the business model component *partners and their roles* is more general than other components, depending on where the link to the respective partner (actor) can be assigned to.

Value Offering Model

The value offering model includes the offerings provided to customers including the value it yields to them. It is the result of value creation/value configuration described in the value creation model. Thus, the value configuration and the partners and their roles describe *how* value is created and *who* participates in this process whereas the value offering model clarifies *what* products and services are offered and which problems the solve.

Hence, the business model component *product/service portfolio* refers to the portfolio of goods and services offered by a company, if applicable broken down into elementary offerings. The offerings are closely coupled to several other business model components. On the one hand, they are allocated to a dedicated customer segment reached via a specific distribution channel. On the other hand, offerings are always connected with a pricing model (which is an element of the profit model).

The *value proposition* outlines the benefits provided to customers, referring to the value creation process that is in turn dependent on available capabilities. That is, this component actually points out the customer needs addressed, or problems selved, respectively. Creating value means thinking in terms of customer benefit – successfully establishing a product or service without addressing a customer problem only works in very few cases.

Importantly, related elements such as the unique selling proposition or success factors of the goods and services offered are excluded from the value offering model. A unique selling proposition is what can be derived from a portfolio that is singular enough to differentiate against competitors – i.e., it is a consequence of the business model, especially the value offering and value creation. Similarly, success factors of a product or service denote what a company should do particularly well to flourish (Weill and Vitale 2001). Therefore, critical success factors should be explicitly addressed in a company's strategy.

Customer Model

The customer model describes the interface to the customer. The *target customers* define the group of customers that is addressed by the service offerings or a specific element of the offering, respectively. Moreover, it outlines the geographic areas to be addressed. A basic decision of a firm is whether to enter the B2C or the B2B sector. Considering the B2C market, different industries can be addressed. Zooming in, each industry offers the possibility to market firms of different size and sophistication (Afuah and Tucci 2003, p. 57). The B2B sector allows for attending several segments that can be defined by demographics, lifestyle, income, and the likes.

The *distribution channel* is closely connected to the target customers, indicating via which channels a product or service is delivered to the customer, connecting the value offering with the dedicated customer (segment). Generally, links can be indirect or direct.

- Direct distribution via a sales force or over a web site (cp. e.g., manufacturer model (Rappa 2001), for instance implemented by Dell,[17] refraining from consulting intermediaries to sell their products).
- Indirect distribution via dedicated mediating actors such as resellers, information or service brokers, or data and process transformation.

Obviously, ICT is an important factor in connection with distribution channels, with progresses in technology continuously providing new opportunities. This is the archetypal example for ICT being enabler and integral part of a business model at the same time.

Although suggested by some authors (e.g., Hamel (2002); Osterwalder (2004)), relationship with customers are not integrated into the business model framework, For instance, customer relationship management (CRM) is a part of a company's marketing activities, exploiting methodologies of customer analysis like surveys, after-sales-management, or complaint management dealing with customer acquisition, customer retention, customer retrieval, add-on selling, and the likes.

Instead, the component *customer integration* needs to be added to describe the integration of customers into value creation processes. Customer integration may range from very low (standardized products), to mass customized goods and services, to real integration of customers as product managers or partners adding value to an overall value proposition. High degrees of customer integration are oftentimes realized by addressing communities rather than individual customers.

As already outlined in Sect. 4.2.4, the competitive landscape is neither a part of the customer model, nor included in any other partial model. Market forces such as price wars or new entrants and other influencing factors, e.g., changes in the demographic structure of target groups or legal issues, are external forces that impact on a business model, but do not belong to business models themselves. Such aspects need to be addressed in the competitive strategy.

---

[17]http://www.dell.com.

Profit Model

Oftentimes referred to as most important element of a business model, the partial model dedicated to generate revenues and profits is certainly a core element explaining the money making logic of a company. On overview on costs and revenues makes it possible to plan the commercial success of a product or service. In order to make assumptions on prices and revenues, it is necessary to know the cost structure of the underlying business. Thereupon, pricing models can be built up which account for revenue streams acquired from different sources.

Just as the partner model, the profit model is transversal, having an influence on all other partial business models. The cost structure is more or less determined by the infrastructure required to create value (i.e., the value creation model and the partner model). Moreover, the prices are offered in connection with the value offering model. Lastly, price discrimination leads to different prices for different target groups.

Special interest to this pillar is paid in Chaps. 5–8. The automated provisioning and composition of Web Services in business networks affects the pricing mechanisms which can be applied in this context and raises some issues which need to be dealt with and which are different from those related to the pricing of conventional products and services.

Generally, the customer is considered the main source of revenue, although models of indirect revenue generation are quite common in e-business. Following these insights, three components categorized into the profit model are introduced, dealing with costs, pricing, and revenues.

The *cost structure* summarizes the expenditures connected with creating (value configuration and transaction and coordination costs incurred by the partner network) and distributing the offered value. The cost structure expresses the relationship between a company's revenues and the underlying costs of generating the revenues (Afuah and Tucci 2003, p. 73). Major cost drivers can be an important indicator when it comes to make-outsource-or-buy decisions.

Pricing models allocate a pricing mechanism to an offered good or service. Generally, there are five types of pricing, which can either be static or dynamic (Afuah and Tucci 2003, p. 60):

- *Menu pricing:* Most common form of pricing with a fixed price offered to customers. For example, such fixed prices can be based on subscription (i.e., flat rate) or on pay-per-use (i.e., metered pricing).
- *One-to-one bargaining:* Negotiation between seller and buyer.
- *Auction:* The seller solicits bids from many buyers and sells to the buyer with the best bid.
- *Reverse auctions:* Sellers compete for business, i.e., place bids on customer requests. In a weakened mode of the reverse auction, buyers propose a price, sellers then decide whether to accept or reject a bid.
- *Barter:* Swapping (goods for goods, goods for services, etc.).

Furthermore, price differentiation strategies can be applied, especially when using static pricing. Price differentiation can be based on several characteristics, for example product features, customer characteristics, volume, or value.

Chapter 5 provides a deeper overview of the different types of price differentiation which are also illustrated by means of some up-to-date examples.

Now that pricing models are known, it is still necessary to determine the way *revenues* are generated and which sources are approached. As an exemplary illustration, Wirtz' distinction among direct vs. indirect revenue generation as well as transaction-dependent vs. transaction-independent revenue generation is considered (Wirtz 2001, p. 85). Transaction-dependent revenues are triggered by the interrelation of the organization and the individual who consumes the organization's respective services. If that is not the case, the revenues are transaction-independent. One refers to direct revenues if the revenues are generated without interaction with a third party. Indirect revenues are earned through third parties such as agents. A special form of such commissions are gain sharing revenues, where an intermediary pockets a share of the savings buyers and/or sellers gain through conduct business via the mediating party (Buchholz and Bach 2001). This implies that it hardly makes sense for business models in electronic business to focus on one revenue model only. Rather, a weighted combination of the various revenue models should lead to an individual hybrid solution. Wirtz (2001) refers to this as a "multi-revenue-stream-optimization." The model of the transaction-independent, indirect revenue generation – data mining revenues in particular – occupy a special role in the Web. Through the branding-up of social software, this is and has been an enormous growth market. It is worth noting that these companies would be of little value without their cumulated user profiles and that they tend to stay in business through data mining revenues and – to a limited extent – through (personalized) banner advertisements (cp. Table 4.4).

The combination of costs and revenues (supported by pricing models), i.e., the profit model, results in the margin structure of the business model. This combination indicates (among others) a business model's sustainability. It is not in the scope of the business model to explicitly list the reasons for a business model to be sustainable. It is the overall business model that answers this issue.

Furthermore, the funding by external capitalists, creditors, or trusters, is not part of rge business model as, for instance, introduced by Wirtz (2001) as "funding model." It is an external factor that is not directly connected with the essence of a company's business, rather being an enabler for conducting it.

Table 4.4 Revenue model

	Direct revenue generation	Indirect revenue generation
Transaction-dependent revenue	Transaction revenues in the narrow sense, connection fees, usage fees	Commissions
Transaction-independent revenue	Setup fees, base fees	Banner advertisement, data mining revenues, sponsorship

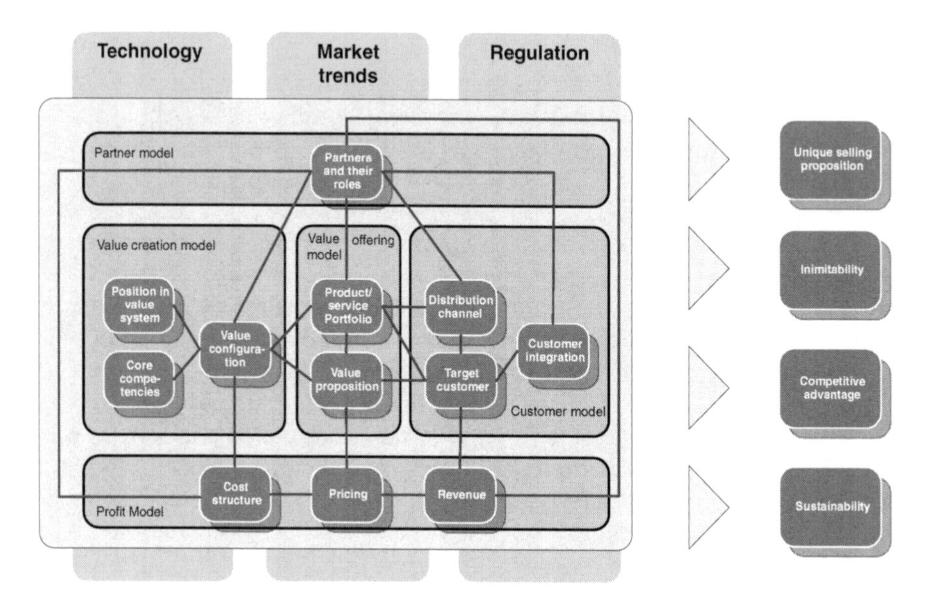

Fig. 4.8 The business model framework and its embedding

4.5 Summary

In conclusion, an overall picture of business models and external influencing factors (trends in technology, market trends, and regulation) was drawn. The very business models itself decides upon crucial questions such as "does the company or value network provide a unique selling proposition?," "is it easy/possible to imitate the business?," "is the business sustainable?," or "does the business generate a competitive advantage?" (cp. Fig. 4.8).

The provision of Web services raises some issues which are different from mere technological aspects. Automatic provisioning and composition of services creates some economic challenges which are related, e.g., to establishing and managing service level agreements within the partner network, to integrating the customer into the value creation, to setting prices and designing incentives in an information environment of numerous and very short-lived and dynamic relationships. These economic issues are covered in a structured way within the concept of the business model in this chapter. This chapter gives an overview of the theory on business models in general, but it also takes into account the peculiarities of business networks for the provision of Web services in order to set up a framework for a business model. It enlists the essential elements which need to be designed in detail and implemented in order to realize successful business activities in a business network of Web services. Pricing as part of the profit model is only briefly mentioned in this chapter, however, a whole part in this book is dedicated to this topic.

References

Abrahamson E, Fairchild G (1999) Management fashion: Lifecycles, triggers, and collective learning processes. Admin Sci Q 44(4):708–740

Afuah A, Tucci C (2003) Internet business models and strategies. McGraw-Hill, Boston

Allee V (2008) Value network analysis and value conversion of tangible and intangible assets. J Intellect Capital 9(1):5–24

Alt R, Zimmermann H (2001) Preface: Introduction to special section – business models. Electron Market 11(1):3–9

Amit R, Zott C (2001) Value creation in E-business. Strat Manag J 22(6–7):493–520

Bach N, Buchholz W, Eichler B (2003) Geschftsmodelle fr Wertschpfungsnetze – Begriffliche und konzeptionelle Grundlagen. Gabler, Wiesbaden

Bamberg G, Coenenberg A (2002) Betriebswirtschaftliche Entscheidungslehre. Vahlen, Munich

Bartelt A, Lamersdorf W (2000) Geschftsmodelle des Electronic Commerce: Modellbildung und Klassifikation. In: Verbundtagung Wirtschaftsinformatik, pp 17–29

Bartelt A, Zirpins C, Lamersdorf W (2001) Geschftsmodelle des Electronic Commerce: Modellbildung und Klassifikation. In: Informatik 2001 – Wirtschaft und Wissenschaft in der Network Economy

Bellman R, Clark C, Malcolm D, Craft F CJ andRicciardi (1957) On the construction of a multi-stage, multi-person business game. Oper Res 5(4):469–503

Betz F (2002) Strategic business models. Eng Manag J 14(1):21–21

Bieger T, Regg-Strm J (2002) Net economy – Die Bedeutung der Gestaltung von Beziehungskonfigurationen. In: Zuknftige Geschftsmodelle: Konzept und Anwendung in der Netzkonomie. Springer, Berlin, pp 15–33

Bieger T, Regg-Strm J, von Rohr T (2002) Strukturen und Anstze einer Gestaltung von Beziehungskonfigurationen– Das Konzept Geschftsmodell. In: Zuknftige Geschftsmodelle: Konzept und Anwendung in der Netzkonomie. Springer, Berlin, pp 35–61

Brugger R (2005) Der IT-BUsiness-Case. Springer, Berlin

Buchholz W, Bach N (2001) The evolution of netsourcing business models: Learning from the past and explore future opportunities. Working Paper 2, Justus–Liebig–Universitt Gieen

Chesbrough H (2003) Open innovation: The new imperative for creating and profiting from technology. Havard Business School Press, Boston

Chesbrough H, Rosenbloom R (2000) The role of the business model in capturing value from innovation: Evidence from Xerox Corporation's Technology Spinoff Companies. Working paper 01–002, Havard Business School, Boston

Chesbrough H, Rosenbloom R (2002) The role of the business model in capturing value from innovation: Evidence from Xerox Corporation's Technology Spin-off Companies. Ind Corp Change 11(3):529–555

Chesbrough H, Vanhaverbeke W, West J (2006) Open innovation: Researching a new paradigm. Oxford University Press, Oxford

Coase R (1937) The nature of the firm. Economica 4(16):386–405

Eisenmann T, Hallowell R, Tripsas M (2001) Internet business models: Texts and cases. McGraw-Hill, New York

Gordijn J, Akkermans H (2001) Ontology-based operators for e-business model de-and reconstruction. In: International conference on knowledge capture, Victoria, BC, pp 60–67

Hagel III J, Singer M (2000) Unbundling the corporation. McKinsey Q 4(3):148–161

Hamel L (2002) Leading the revolution: How to thrive in turbulent times by making innovation a way of life. Harvard Business School Press, Boston

Heinrich B (2000) Dimensionen zur Beschreibung eines Geschftsmodells fr Kreditinstituteim Bereich Privatkunden. Working paper, St. Gallen

Heinrich B, Leist S (2000) Nutzung und Entwicklung von Geschftsmodellen – Ergebnisse desKompetenzzentrums Bankenarchitekturen im Informationszeitalter. In: Business Engineering: Auf dem Weg zum Unternehmen des Informationszeitalters. Springer, Berlin, pp 329–407

Kagermann H, sterle H (2006) Geschftsmodelle 2010: Wie CEOs Unternehmen transformieren. Frankfurter Allg. Buch, Frankfurt

Klueber R (2000) Business model design and implementation for eservices. In: Americas conference on information systems (AMCIS), pp 797–800

Kotler P, Armstrong G (2001) Principles of marketing. Prentice Hall, Upper Saddle River

Lambert S (2006) Do we need a "Real" taxonomy of e-business models?. School of Commerce Research Paper Series 06–6

Linder S, Cantrell J (2000) Changing business models: Surveying the landscape. Working paper, Accenture Institute for Strategic Change

Lindström C (1999) Lessons learned from applying business modelling: Exploring opportunitiesand avoiding pitfalls. In: Perspectives on business modelling: Understanding and changing organisations. Springer, Berlin, pp 151–164

Magretta J (2002) Why business models matter. Harv Bus Rev 80(5):86–93

Mahadevan B (2000) Business models for internet-based e-commerce: An anatomy. Calif Manag Rev 42(4):55–69

Mintzberg H (1987) The strategy concept I: Five Ps for strategy. Calif Manag Rev 30(1):11–24

Mller-Stewens G, Lechner C (2001) Strategisches management. Schffer–Poeschel, Stuttgart

Nilsson A, Tolis K, Nellborn C (1999) Perspectives on business modelling: Understanding and changing organisations. Springer, Berlin

O'Reilly T (2005) What is Web 2.0. http://www.oreilly.com/pub/a/oreilly/tim/news/2005/09/30/what-is-web-20.html, Accessed on 28 Nov 2007

Osterwalder A (2004) The business model ontology – A proposition in a design science approach. PhD thesis, L'Ecole des Hautes Etudes Commerciales de l'Universit de Lausanne

Osterwalder A, Pigneur Y (2002) An e-business model ontology for modeling e-business. In: 15th blend electronic commerce conference – e-reality: Constructing the eEconomy, pp 75–91

Osterwalder A, Pigneur Y, Tucci C (2005) Clarifying business models: Origins, present and future of the concept. Commun Assoc Inform Systems 16(1):1–25

Papakiriakopoulos D, Poulymenakou A, Doukidis G (2001) Building e-business models: An analytical framework and development guidelines. In: 14th Bled electronic commerce conference, pp 446–464

Pateli A, Giaglis G (2003) A framework for understanding and analysing e-business models. In: 16th Bled ecommerce conference etransformation, pp 329–348

Petrovic O, Kittl C, Teksten RD (2001) Developing Business Models for eBusiness, Proceedings of the International Conference on Electronic Commerce, Vienna, Austria

Pohle G, Korsten P, Ramamurthy S, Foecking S (2005) The specialized enterprise: A fundamental redesign of firms andindustries. IBM Business Consulting Services

Porter M (1985) Competitive advantage. Free Press, New York

Porter M (2001) Strategy and the internet. Harv Bus Rev 79(3):62–79

Rappa M (2001) Business models on the web. http://digitalenterprise.org/models/models.html, Accessed on 16 Jan 2008

Sandrock J, Weinhardt C (2006) Geschftsmodelle fr E-Learning Anbieter: ein Framework fr dieStrategische Planung. Zeitschrift fr Betriebswirtschaft Special Issue (2):18–41

Scheer C, Deelmann T, Loos P (2003) Geschftsmodelle und internetbasierte Geschftsmodelle Begriffsbestimmung und Teilnehmermodell. Working paper, Reasearch Group Information Systems and Management

Schögel K (2002) Bezugsrahmen der Geschftsmodellierung. In: Roadm@p to E-Business. Thexis, St. Gallen, pp 374–399

Schwickert A (2004) Geschftsmodelle im Electronic Business - Bestandsaufnahme undRelativierung. Working paper, Justus-Liebig-Universitt Gieen

Stathel S, Finzen J, Riedl C, May N (2008) Service innovation in business value networks. In: Proceedings of XVIII international RESER conference, pp 288–302

Sthler P (2002) Geschftsmodelle in der digitalen konomie. Josef Eul Verlag, Lohmar

Tapscott D, Williams A (2006) Wikinomics: How mass collaboration changes everything. Portfolio, New York

Tapscott D, Lowy A, Ticoll D (2000) Digital capital: Harnessing the power of business webs. Harvard Business School Press, Boston

Timmers P (1998) Business models for electronic markets. EM – Electron Markets 8(2):3–8

Timmers P (1999) Electronic commerce. Wiley, New York

Toffler A (1980) Future shock: The third wave. Bantam Books, New York

Weill P, Vitale M (2001) Place to space: Migrating to ebusiness models. Harvard Business School Press, Boston

Wirtz B (2001) Electronic business. Gabler, Wiesbaden

Yu C (2001) An integrated framework of business models for guiding electronic commerce applications and case studies. Electron Commerce Web Technol pp 111–120

Zimmermann H (2000) Understanding the digital economy: Challenges for new business models. In: Americas conference on information systems (AMCIS), pp 729–732

Part II
Web Service Pricing

Chapter 5
Pricing Foundations and Implications on Web Service Pricing

5.1 Pricing Principles in General

One of the key building parts of a market is its market mechanism. This mechanism encloses the rules for allocation and pricing and thus regulates and, in some cases, enforces the procedures of trading on a specific market. In general, both issues of allocation and pricing are intrinsically tied to one another. Neumann et al. (2007) and Buyya et al. (2008) treat online trading platforms. For example, such as Grid and Cloud service exchanges. Yet, since allocation procedures are often directly connected to technical conditions, in the following the focus is mostly on the pricing mechanisms that are used in today's markets. Basically, one can distinguish between static, flexible, and dynamic pricing, which can be further subdivided into concrete pricing schemes.

5.1.1 *Static Pricing*

Static pricing can be regarded to be the simplest form of pricing mechanisms. The price of a unit of a good is constant and is also referred to as linear (or uniform) because the amount of money that is paid by the customer is linearly related to the volume of the good purchased by this same customer. Naturally, the price depends on the costs for the provision of the good. These costs are usually made up of fixed costs, that is, the initial investment to achieve the capability to produce the good and which costs are independent of the volume produced eventually, as well as variable costs that are directly linked to the quantity produced. The ratio between fixed and variable costs can vary a lot depending on the kind of product. While for simple physical goods sunk fixed costs may be low and main costs occur during the production process itself. Shapiro and Varian (1999) note that for information goods the contrary holds, that is, the sunk fixed costs to initially

C. Weinhardt et al., *Business Aspects of Web Services*,
DOI 10.1007/978-3-642-22447-8_5, © Springer-Verlag Berlin Heidelberg 2011

produce an information good are much larger than the costs for reproducing this information.

Even though in a competitive environment costs are the main factor for the pricing decision of a single firm, the demand for the good plays also a role which is the more important from the perspective of the single supplier the more concentrated the market is. The mark-up (the difference between the price and the costs) will be higher the more market power the supplier has and the more inelastic is the demand the supplier faces. In particular, the price–cost margin of a monopolistic firm is inversely related to the price elasticity of demand (Tirole 1988; Gravelle and Rees 1993). This is illustrated in Fig. 5.1 where a monopolist with decreasing returns to scale (the average cost curve is decreasing and lies below the marginal cost curve) sets the profit-maximizing price for a good. The optimal price P^* is found at the (inverse) demand curve ($P(q)$) for a quantity q^* which equalizes marginal revenue *MR* and marginal costs *MC*. The realized profit is the area enclosed by q^* and the difference between the price P^* and average costs $C = AC(q^*)$. One can easily verify that this area would be larger for a steeper (more inelastic) inverse demand curve.

Even if the firm is able to set as high a price as the monopoly price from the consumers, the static pricing strategy is usually considered to be less profitable and not promising to achieve comparably high revenues as with advanced pricing schemes.

This notion is based on the fact that static pricing does not take into account other information that probably is available like, for example, differences in customers' preferences. The flexible pricing schemes that will be discussed in the next section were designed to make use of such information in order to extract more surplus from the consumers.

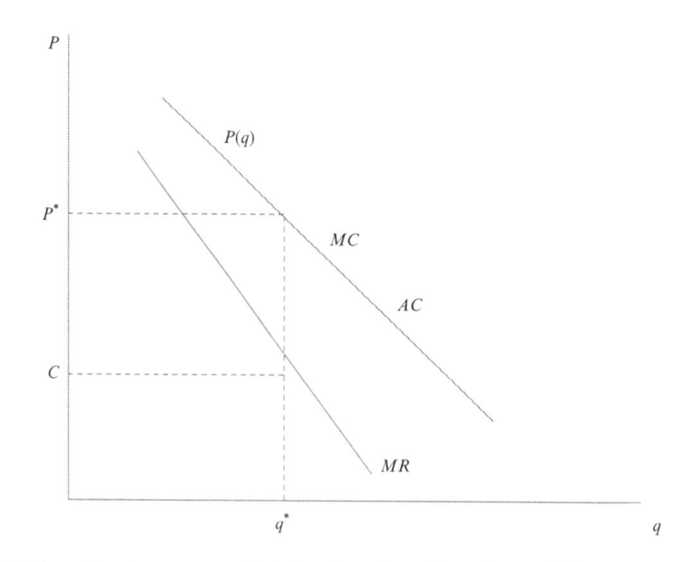

Fig. 5.1 Static pricing by a monopolistic firm (see Gravelle and Rees (1993), p. 272)

5.1.2 *Flexible Pricing*

Flexible pricing mechanisms are designed to take into account specific customers' preferences and make use of them in order to achieve higher sales and revenues. Products that are designed to meet customers' requirements (up to a certain degree) enable the customers to choose that product from a variety of versions which fits their needs best. For the supplier, the probability of selling a certain product (or product version) rises, since, given a larger product set to choose from, the customer may acquire a product he would otherwise not purchase at all (as, for a given price, the product's properties, including the price, differ too much from the customer's preferences). Speaking of customer's preferences may refer to several aspects, such as the quality of the product (or, for instance, the quality of service of a Web service) or product related features. Differences in the preferences of customers allow firms to increase their profits by product differentiation and price discrimination. However, in order to present the matter in a compact fashion, this chapter focuses on the customers' willingness to pay for different product offers. Firms may then engage in forms of either price discrimination or product differentiation with varying prices in order to increase their profits.

Flexible pricing is realized by price discrimination.[1] The method of price discrimination to be employed depends on the available information about customers that are expected for the specific product as well as on the possibilities of arbitrage in the market(s). Arbitrage between customers is based on transferability of the product, which applies only to a limited degree for services by their nature. The more detailed the available information is and the less trading activities customers can engage in, the better can the market be skimmed by the provider. Price discrimination is structured into three different degrees.

In this section, the basic ideas of price discrimination and several mechanisms that implement these are outlined. A general overview is given by Tirole (1988). For a more detailed discussion, please refer to Shapiro and Varian (1999); Varian (1997). The figures in this section also follow their style.

5.1.2.1 Personalized Pricing

First degree price discrimination is applied in the case that there is perfect information about each customer's willingness to pay. A producer (or service provider) will offer the product (or service, respectively) at a price that is equal to or marginally lower than the respective willingness to pay of the consumer under the assumption that there is no collusion among customers. Thus, the perfect information available to the provider is completely exploited and maximizes the provider's revenue. The difference in profits between simple static pricing and personalized pricing

[1] Also: price differentiation.

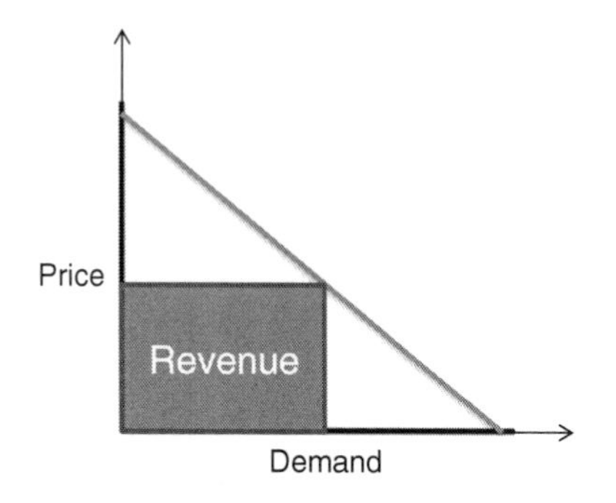

Fig. 5.2 Provider's revenue with static pricing

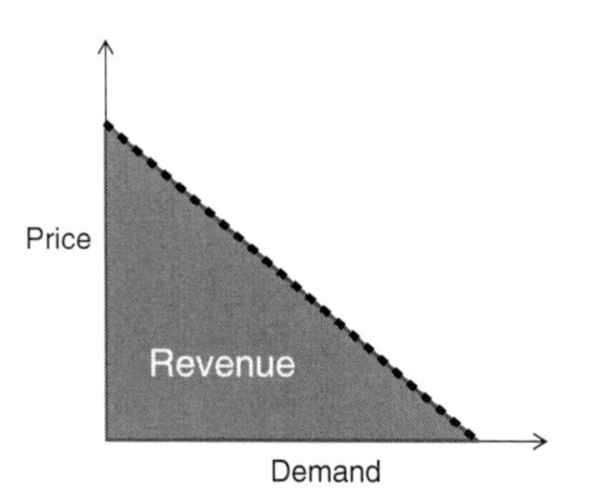

Fig. 5.3 Provider's revenue with personalized pricing

is illustrated in Figs. 5.2 and 5.3. While for static pricing with a fixed price p all customers with a willingness to pay higher than p buy the product at this price, customers with a lower willingness to pay do not purchase the product at all. This means potential losses on both sides, as can be seen in Fig. 5.2. This drawback is eliminated by personalized pricing, as seen in Fig. 5.3. However, personalized pricing is a merely theoretical pricing scheme and denotes the ideal case (from the point of view of the provider). Since in real scenarios it is extremely challenging for a provider to determine each individual's willingness to pay, such a scheme can hardly be enforced and is, therefore, impractical. This approach becomes even more complicated when one considers that in many cases not even the consumers can estimate their own willingness to pay. The pricing mechanisms that are introduced in the following section serve to achieve an applicable solution while sacrificing the theoretical profit-optimizing solution.

5.1.2.2 Self-Selection: Versioning, Bundling, and Amount-Based Pricing

Price discrimination of second degree is a practically feasible approach that provides the seller with a means to achieve higher revenues as compared to static pricing by

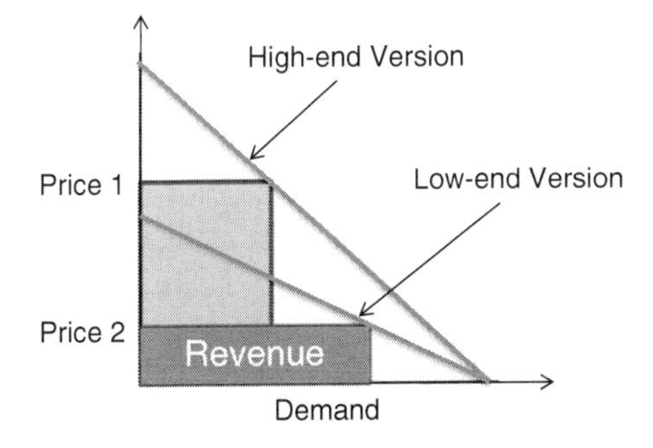

Fig. 5.4 Provider's revenue with versioning

offering his product in different ways. The pricing schemes discussed in this section are versioning, bundling, and amount-based pricing. As these pricing schemes usually leave the customer with a choice to select from several varieties bundles of the product, price discrimination of second degree is also called *self-selection*.

When applying versioning, a provider offers a product in different versions at different prices, for example, a high-end version at a high price and a low-end version at a low price. While the offer in Fig. 5.2 could be seen as a single version only, in Fig. 5.4 the provision of two versions of a product, high-end and low-end, is illustrated exemplarily. In the versioning scenario, customers with a lower willingness to pay will acquire the latter, while customers with a higher willingness to pay or the necessity for specific features will buy the high-end version. For the customer preferring the high-end version, the price savings from the low-end version are not sufficient to offset the missing product features. Versioning of a product is usually facilitated by a variation in price, quality of the product, or specific features. A simple example would be an internet service provider, offering several products (i.e., access to the internet) that differ in speed and the provided support.

The challenge of offering different versions of one product is to determine what kind of versions to offer at which price. At first glance, this problem might seem similar to the personalized pricing approach. Nevertheless, versioning is much more applicable as it does not require to specify prices according to individuals' willingness to pay but rather prices for a limited amount of product versions. Since the focus of versioning is on the product itself, market surveys can usually be used to estimate the appropriate version and price combinations. Even though in many cases it still remains a challenging task to design the optimal product versions and their according prices to achieve the maximal revenue.

While versioning sets different prices for different versions of one product in order to approximate the optimal revenue rates of personal pricing, bundling aims to achieve that by offering several products in one package. Therefore, the customers may choose whether to acquire products separately or in a bundle. This approach is particularly useful, when it is assumed that the customers have heterogenous willingness to pay for each product. The following examples illustrate this.[2]

Example 5.1. There are two different products p_1 and p_2 and two customers c_1 and c_2, with heterogeneous willingness to pay for each product. While c_1 is willing to pay 8 and 4 monetary units for p_1 and p_2, respectively, customer c_2 has the opposite willingness to pay vector, that is, 4 and 8 monetary units for products p_1 and p_2. In case the marginal costs for the product provisioning are assumed to be zero, it can be shown that for a (monopolistic) provider it is optimal to set the prices for each product to 8, and to charge 12 for the bundle. In this way, the bundle always turns out to be profitable, as both customers are willing to acquire the bundle, rendering 24 monetary units of profit. If p_1 and p_2 would be offered separately only, c_1 would only by p_1, and c_2 only p_2, yielding a profit of only 16 monetary units.

The reason for the profitable usage of bundling is that it effectively reduces the dispersion of the customers' willingness to pay. One can distinguish between two strategies, pure and mixed bundling. Pure bundling means to offer different products only in a bundle and not separately, while mixed bundling allows both. In this way, mixed bundling enables to discriminate between different consumer groups by providing a set of alternative offers. The potential benefit of this approach is shown in the following example.

Example 5.2. There are two products p_1 and p_2 and four customers $c_1, \ldots, c_4$. The willingness to pay vectors for the customers are given by $\{1, 9\}$, $\{4, 8\}$, $\{8, 4\}$, $\{9, 1\}$, respectively. In case the products are offered separately, it is optimal for the provider to set both prices at 8, giving him an eventual profit of $4 \cdot 8 = 32$. Employing pure bundling, the revenue is maximized by charging for the $\{p_1, p_2\}$ bundle 10 monetary units, thus, rendering 40 units for the provider through sales, as all customers will buy the bundle. However, by following a mixed bundling strategy, and setting the price for the bundle to 12 and for each product to 9 units, will render a profit of $2 \cdot 9 + 2 \cdot 12 = 42$ units – the first and last customer will buy the one product only, while the second and third customer will buy the bundle.

Amount-based pricing is usually applied when the product to be sold involves certain quantities. Examples would be the provision of a household with water, electricity and gas, or providing access to the internet, which may be based on time or the quantity of transmitted data. In this case, the provider of the respective good or service may offer the consumer to choose, for example, between a usage-based pricing (that is, fixed per quantity of used unit) or a fixed price independently of this amount, that is, a flat fee. Other amount-based pricing schemes would be to set a varying price depending on the quantity used (i.e., decreasing with a higher amount

[2]Note that these examples implicitly state the assumption that each customer's willingness to pay for a bundle of products is the sum of the willingness to pay for each single product contained in this bundle.

of used units) or combinations of usage-based pricings and flat fees (i.e., after a certain cap of used units is reached, the pricing scheme switches to a flat fee). For a more detailed discussion, see Sundararajan (2004). In general, it can be said, that flat fees are particularly useful for customers that consume large quantities of units, as therefore the average price per unit eventually approaches zero. On the other hand, it seems obvious that users with only a low consumption generally should favor usage-based pricing. However, it has been shown that even this kind of consumers often prefers flat fees, though from an economic point of view they are not optimal for them. In the recent literature, this phenomenon is also called *tariff bias* and is at least partially attributed to the so-called *taximeter effect*. For a deeper discussion of this topic see, for example, Lambrecht and Skiera (2006) and Krämer (2010).

In addition to the above pointed out pure pricing schemes for first and second degree price discrimination, there exist also various forms of pricing that cannot be assigned exclusively to one of these. One example of this is given by reverse pricing (see Chernev (2003)). The provider of a product has a secret price limit for which he is willing to sell his product. Each customer is allowed to state a bid, that is, how much he is willing to pay. If this bid is above the provider's price limit, the customer pays the amount according to his bid. Since successful bids are not announced to other customers than the original bidder, there is no price competition among the customers. Reverse pricing is thought of in-between static pricing (with an unknown fixed price) and personalized pricing. A market that actually implements this pricing scheme is http://www.priceline.com/.

5.1.2.3 Group Pricing

First degree price discrimination charges each customer individually (under perfect knowledge of the customers' preferences) and, thus, maximizes the (theoretical) profit. Since this is not feasible in practice, second degree price discrimination tries to avoid this complex problem by offering certain product alternatives (by versioning, bundling, or different billing schemes) and letting the customers choose by themselves what is optimal for them. Thus, discrimination takes place by customers' self-selection by means of their own choice between different products or tariffs. Third degree price discrimination, in contrast to second degree price discrimination, focuses on the customers again by directly using some observable characteristic of the customers. For instance, group pricing is a form of third degree price discrimination and approximates the idea of personalized pricing by subdividing customers not individually but rather into groups and, thus, rendering it applicable in real world scenarios. The challenge when applying group pricing is to subdivide the customers into groups that are robust and can be verified. Examples would be students, senior citizens, gender or citizenship biased, depending on the kind of product for sell.

Comparing Figs. 5.4 and 5.5 one notes the similarity between the realized profits achieved by both approaches, i.e., versioning and group pricing. The crucial difference is that versioning involves the customers' self-selection aspect,

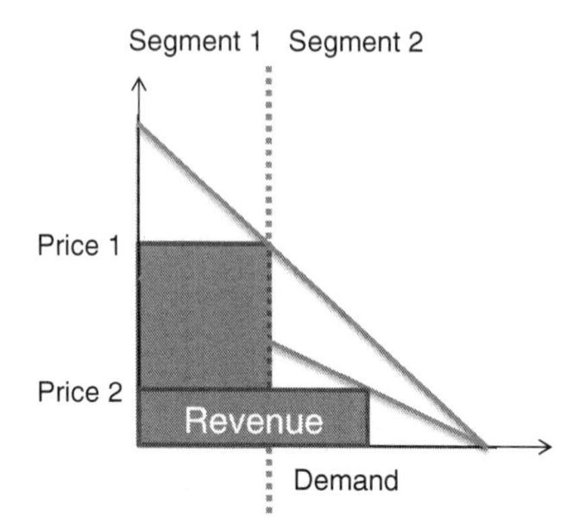

Fig. 5.5 Provider's revenue with group pricing

while in group pricing the customers' choice is limited by the group they belong to. Furthermore, while versioning works by varying product quality and additional features, group pricing usually is applied by (but is not limited to) monetary differentiation. It must be noted that both approaches can also be combined to even more complex pricing schemes, that is, offering specific versions only to a certain group of customers.

5.1.3 Dynamic Pricing

Dynamic pricing, according to Wikipedia, is applied in numerous fields. The general notion of dynamic pricing is that the product prices are determined according to usually complex dynamic pricing schemes that vary in dependency to several variables or additional conditions. These pricing schemes provide more flexibility than the above named static and flexible pricing approaches and, thus, enable the market participants to adapt better to actual market situations trying to maximize their profit or minimize their expenses. Albeit the more profound complexity of flexible over static pricing mechanisms, these approaches still do not enable any trader to systematically adapt their behavior to fast changing market situations. Dynamic pricing just offers providers as well as customers more options to act. However, this richer strategic space in dynamic pricing schemes is more challenging to control, that is, in the sense to, for example, find the (theoretically) optimal trading strategy. Usually, both providers and customers are involved actively in determining the outcome, or one side uses at least their expectation about the other side's behavior. It can be noted that time is often a crucial factor among the parameters used in the pricing formulas. This section particularly focuses on auctions as one kind of dynamic pricing. Other dynamic pricing schemes are presented in Chaps. 6 and 8.

Auctions are one of the most prominent example of dynamic pricing schemes. Depending on the specific auction type, both sides of the market are actively involved (for example, by stating their bids or reservation prices) and, additionally, are often in competition to other providers or consumers, respectively. Today the most distinct auction forms have been developed and established successfully. For many of these auctions optimal (bidding) behavior (given certain assumptions) can be shown analytically. The type of auction that is employed usually depends on the goods (and their characteristics) that are to be sold via the auction, often also involving historical and traditional aspects. Further readings about auctions, their characteristics and practical employment as well as theoretical aspects are given by Klemperer (1999), Krishna (2009), and Milgrom (2004). Note that auctions, in contrast to the mechanisms presented above, comprise pricing as well as allocation rules. The four most common auction types are the English, Dutch, first-price sealed-bid, and Vickrey auction.

In an English auction several customers compete for the same product (or set of products). Each bidder may state his bid which must be higher than the last bid that was stated openly by another competing consumer. Each bidder may adapt his bid repeatedly according to the other bidders' behavior. The bidders will leave the auction as soon as the price rises above their willingness to pay, which, in contrast to their stated bids, remains unknown to the other bidders. The last bidder remaining in the auction process is declared as the winner and pays his last bid as the price to acquire the product. English auctions are probably the most common auction form for selling paintings, antiquities or other assets whose value can hardly be estimated and is subject to a large dispersion among the bidders willingness to pay.

In a Dutch auction, the auctioneer publicly announces the current price for the good to be acquired. This price usually starts with a high reservation price and is decreased in priorly announced fixed steps until the first bidder accepts the price. This bidder wins the auction and pays the price that he accepted. Contrary to the English auction, the bidders in the Dutch auction can only state an equivalent of their bid by deciding how much time to wait until committing themselves to pay the actual price for the good. In this way, English auctions, where no fixed price interval for the next higher bid is set, may enable the bidders to more strategic behavior, for example, by raising the bid by an unusual high amount, thus, creating and exploiting psychological effects. Dutch auctions are mostly known for their application in Dutch flower selling markets but also have recently been established in online markets, for example, http://www.1-2-3.tv/.

In contrast to the sequential format of the English and Dutch auctions that allow increasing or decreasing bids, respectively, first-price sealed-bid auctions allow the bidders to submit just one bid. This bid is sealed and hence, only known to the bidder. It is stated independently from other participants' bids and thus, the procedure avoids any direct competition. It can be shown that in this auction form the participating consumers will not bid their true willingness to pay, but tend to understate it with their bid (see, e.g., Ausubel et al. (1996), Zeithammer (2007) and the references therein). This so called *bid shading* is also the result of rational behavior in *Nash equlibrium*[3] for first-price sealed-bid auctions. Prominent examples are, for example, (public) tenders.

The disadvantage of the first-price sealed bid auction is that no dominant strategies exist. While, for example, in the English auction it is clearly profitable for a customer to stay in the auction as long as the actual price is below the auction participant's willingness to pay, no such strategies exist neither for the Dutch nor for the first-price sealed-bid auction. A different approach is employed in the Vickrey auction that is carried out analogously to a first-price sealed-bid auction, with the exception of the price determination. Instead of paying his own bid for the good, the winning participant is obliged to pay the second highest bid. Therefore, the Vickrey auction is also called second-price sealed-bid auction.[4] For this auction form it can be shown that it is a dominant strategy for each bidder to bid his own valuation, i.e., willingness to pay. This strategy is also known as *truth telling*.

A generalized format of the Vickrey auction is a multi-unit auction. In this case, bidders submit bids for single units as well as available bundles. The winner determination is carried out similarly to Vickrey auctions but requires the distribution of all available items. Hence, there can be single bidders that win with a highest bid for all of the items, or several bidders of which the (sealed) bids jointly form the highest bid. Price determination is carried out separately for each of the winning bidders. In order to calculate the price to be paid by bidder i, the maximum bid when removing the participant i's bids is reduced by the originally winning bids without bidder i's bids. The resulting amount reflects a theoretical second highest price.

The question that might now arise for a provider is via which auction form he should offer his products in order to achieve the (theoretically) highest revenue. Though no answer can be given in general, it can be stated that under certain mild assumptions about the bidders and their characteristics (see McAfee and McMillan (1987) for details) all the above four illustrated auctions yield the same outcome. This is also known as the *Revenue equivalence theorem*.

The above named auction types can be characterized according to the following attributes. Since in all depicted auctions only the consumers take an active role as bidders, the auctions are characterized as being *one-sided*. Double-sided auctions follow more complex bidding rules, involving the consumers as well as the providers stating their bids. Further characteristics are whether the bids are made public (i.e., open) or sealed, and whether these bids are stated in an ascending or descending order. This classification is illustrated in Fig. 5.6. Therefore, an English auction is an open ascending auction, while a Dutch auction is open but descending. Both the first-price sealed-bid as well as the Vickrey auction are one-sided, sealed auctions. Examples for double sided auctions are Call-markets (with sealed bids), as they occur at the start of each trading day at stock exchanges, and continuous double auctions (with open bids), which manage the ongoing trade in these exchanges.

---

[3] That is, no participating party can achieve a higher utility by unilaterally deviating from her bid determined in the equilibrium.

[4] nth-price sealed-bid auctions can be defined analog.

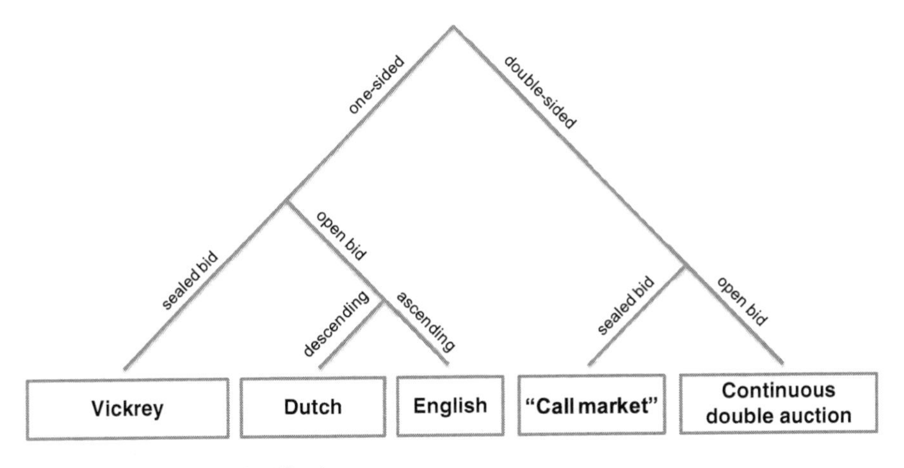

Fig. 5.6 Auction types classification

5.2 The Pricing of Web Services

The question that now arises is whether and how the pricing mechanisms presented in the previous section can be used for the pricing of Web services. The answer to this question must take into account the specific characteristics of Web services pointed out in Chap. 4 as well as the needs that may differ when customers acquire Web services instead of ordinary goods.

Static pricing, though in most cases not the optimal choice for the provider, can be applied as a pricing scheme for almost any goods, including Web services. For first degree price discrimination there are the same challenges for their application in Web service pricing as for other goods, that is, it remains a merely theoretical approach. As versioning and bundling has already successfully been employed with software sales, there is no constraint regarding its application in Web service markets. A similar argument holds for amount-based pricing. Therefore, for second degree pricing schemes, there is no restriction regarding their usage in pricing Web services.

Group pricing is also a promising approach to be employed within the sales of Web services. In traditional group pricing scenarios, group segmentation and the verification of the affiliation of customers may be challenging (or, at least, not automated) in particular cases. The advantage of applying third degree price discrimination in Web service markets is that Web service providers may use the already established infrastructure (including authentication and authorization) that exist for virtual organizations (see, e.g., Foster et al. (2001)).

There are no apparent arguments against the application of any of the flexible pricing schemes to Web service pricing. This, however, cannot be said automatically for dynamic pricing, for which specific aspects of Web services and their related trading need to be taken into account. As Web services form often part of some automated processes (in, e.g., Service Value Networks, illustrated in Chap. 3) the sometimes time consuming price and allocation mechanisms of the above depicted auctions (e.g., the English and Dutch auctions) might hinder the application of these schemes in Web service markets. Even if one considers usage of automated bidding processes and agents (see, e.g., Borissov (2009)), this only diminishes the dimension of this problem. Also, introducing the option to buy auctioned Web services instantaneously is not satisfying. This immediate purchase option, while bypassing the regular auction procedure, will cost the consumer a posted price that will usually be above the expected price the Web service would yield in the regular auction process, thus, denoting an additional charge. For Web services whose usage can be planned in advance, the implementation of auctions as Web service market mechanisms states no contradiction. The above named restrictions do not hold for all dynamic pricing schemes, of course, as will be illustrated in the following Chaps. 6 and 8.

References

Ausubel L, Cramton P (1996) Demand reduction and inefficiency in multi-unit auctions. University of Maryland, College Park

Borissov N (2009) Engineering strategies for market-based scheduling of computing services. In: Fourth International Workshop on Massively Multi-Agent Systems: Models, Methods and Tools (MMAS), in conjunction with AAMAS 2009

Buyya R, Yeo C, Venugopal S (2008) Market-oriented cloud computing: Vision, hype, and reality for delivering it services as computing utilities. In: high performance computing and communications, 2008. HPCC'08. 10th IEEE international conference on IEEE, pp 5–13

Chernev A (2003) Reverse pricing and online price elicitation strategies in consumer choice. J Consum Psychol 13(1):51–62

Foster I, Kesselman C, Tuecke S (2001) The anatomy of the grid: Enabling scalable virtual organizations. Int J High Perform Comput Appl 15(3):200

Gravelle H, Rees R (1993) Microeconomics. Longman, Harlow

Klemperer P (1999) Auction theory: A guide to the literature. J Econ Surv 13(3):227–286

Krishna V (2009) Auction theory. Academic, Salt Lake City

Krämer LJ, Wiewiorra (2010) Consumer perception of mobile telephony tariffs with cost caps. In: Proceedings of 9th conference on telecommunications, internet and media techno economics (CTTE), IEEE Xplore, Ghent, Belgium, pp 1–8

Lambrecht A, Skiera B (2006) Paying too much and being happy about it: Existence, causes, and consequences of tariff-choice biases. J Market Res 43(2):212–223

McAfee R, McMillan J (1987) Auctions and bidding. J Econ Lit 25(2):699–738

Milgrom P (2004) Putting auction theory to work. Cambridge University Press, Cambridge

Neumann D, Stoesser J, Anandasivam A, Borissov N (2007) Sorma-building an open grid market for grid resource allocation. In: Proceedings of the 4th international conference on Grid economics and business models. Springer, Berlin, pp 194–200

Shapiro C, Varian H (1999) Information rules: A strategic guide to the network economy. Harvard Business Press, Boston

Sundararajan A (2004) Nonlinear pricing of information goods. Manag Sci 50(12):1660–1673

Tirole J (1988) The theory of industrial organization. MIT, Cambridge

Varian H (1997) Versioning information goods. University of California, Berkeley

Zeithammer R (2007) Research note strategic bid-shading and sequential auctioning with learning from past prices. Manag Sci 53(9):1510

Chapter 6
Pricing Strategies for Platform Providers

6.1 Introduction

For several years, a change in the economy has been observed. The established focus on products made way for a new perspective that concentrates on the provisioning rather than only the result, that is, on services. This development affects technical as well as organizational aspects of the economy. On the one hand, service mashups and situational approaches are facilitated by dynamically composing Web services. RESTful architectures (Fielding and Taylor 2002) and slim messaging formats like JSON (Crockford 2006) support the technical feasibility of the aforementioned composition. This way, the interplay of numerous service providers is enabled which creates value by integrating the provided modules into one joint solution, a complex service that is suited to meet individual customers' requirements.

These complex services typically comprise business functionality that can be seperated into numerous steps (Papazoglou 2008). This is facilitated by the assembly and invocation of service modules, which are offered by diverse service providers that leverage their core competencies in Service Value Networks (SVNs) in order to face customer requirements. Once a customer requests a particular service, service modules are dynamically composed from a pool of complementary service offerings. In summary, these trends foster a rapid growth of SVNs (Blau et al. 2009) that are formed in a short-term fashion in order to provide customized complex services, i.e., composite services that provide added business value, to multifaceted service consumers.

Blau et al. (2009) provide an example of a complex service by means of a payment processing service which is requested by an end consumer. It is composed successively of single service components such as data verification of the customer's address and credit card account followed by a transaction processing service charging the customer's account which is itself followed by a database service updating the change in the customer's account and finalized by a storage service that stores the changes to the data. Accordingly, providers of the single service components

C. Weinhardt et al., *Business Aspects of Web Services*,
DOI 10.1007/978-3-642-22447-8_6,

are required to enter into SLAs with their counterparty at the preceding and the subsequent value-added step.

Economic aspects of SVNs do not only comprise value creation, which can take place in a successive manner along the value chain in terms of the provision of the single service components, but also the establishment of the required SLAs, which govern the functional as well as non-functional aspects of service provisioning, as well as prices and penalties, which are applied in case of a violation. Penalties are a major part of SLAs. First, penalties allow to handle the uncertainty which is inherent to service delivery by compensating a service customer in case of SLA violation. Second, penalties enable the incentivation of service providers (cf. Becker et al. (2008)).

This chapter's focus is on pricing and penalty design decisions once particular service providers in a service value network have decided to enter into SLAs with each other and not – as in other chapters – on the optimal choice of service offers.

The decision on the design of penalty and price in an SLA depends immensly on the decision-maker's expectation on the likelihood of the application of a penalty and, consequently, the delivered quality of service (QoS). The delivered QoS is influenced by the service provider, to a certain degree by the customer's behavior, and to some extent by factors, which are beyond the control of the service provider. Nevertheless, it is reasonale to assume that an increasing effort carried out by the provider leads to a decreasing risk of service failure of a single service and consequently, to a reduced risk of service failure of the complex service, and a lower likelihood of violating the SLA.

The service provider's effort in this context comprises any endeavors or expenses which increase the probability of service delivery. Examples for effort include costs related to hardware exchanges, changes in backup routines, infrastructure related to power generation, personnel, briefly anything the provider could do to influence service provision. In the model in this chapter the effort level, which does not need to have a monetary nature in itself, is monetized by means of the marginal costs of effort to enter the profit function of the providers. In general, these marginal costs might differ across the service providers to allow for heterogeneity. As a special case, effort might be interpreted to represent directly the monetary value of a financial investment which providers make to improve the probability of service delivery. In that case, the variable used in the model to denote the marginal costs of effort will be equal to 1 for all service providers.

Even if observable to the contracting party, the service providers' level of effort is not subject of the SLAs and not enforceable. This is why penalties are not tied to a service provider's failing to exert the required level of effort but to his failing the service level agreement. Thus, the pricing decisions in the SVN, which also include the corresponding penalties for violating the SLA, indirectly incentivize service providers to exert the necessary effort for the provision of their single services. In general and for given prices, higher penalties will induce a service provider to exert more effort to provide the agreed upon service.

This chapter investigates the resulting prices and penalties in the case that service providers compete for service provisioning of complementary service modules, which are integrated into one service. Contrary to works which focus on the efficient allocation of services this work assumes that the probability of service delivery is influenced by a provider's effort and, hence, is endogenous rather than taking the probability of service delivery as some immanent property. The implications that result from the model at hand can be applied to more general scenarios and, hence, can support real-world decision-makers to find efficient levels of prices and penalties that reflect changes in the environment.

The chapter provides a model representing the entities establishing contracts with each other in a service value network. The model comprehends a service consumer and service providers competing for the provision of complementary single services to form a complex service. In an environment of perfect competition at each value-added step, bargaining between the service providers is simplified in that the service provider at each step sets the price and penalty for the provider at the next step. The optimal prices and penalties are analytically derived and comparative statics analysis is applied to show the impact of the variables that are exogenous to the model on the effort level of service providers and on the resulting penalties.

The chapter is structured as follows. Section 6.2 gives an overview of the related literature. In Sect. 6.3, the model and assumptions are presented. Section 6.4 deals with the service providers' behavior once they have established a service level agreement with the consumer and each other. This section shows how they choose, the level of effort for given prices and penalties. Section 6.5 derives the service providers' optimal prices and penalties. In Sect. 6.6, comparative statics analysis is performed to show the effect of exogenous shocks on the choice variables. Finally, Sect. 6.7 discusses the results, especially in view of strategic recommendations, but also comments the limitations of the model and gives an outlook for further research.

6.2 Related Literature

The composition of loosely-coupled service modules into customized complex services on platforms like Salesforce[1] has to be facilitated by means of a common understanding of the description of and the interface to each of these service modules. This standardization is applied in service oriented architectures (SOAs), which allow for reusability and integration in different branches (Leymann 2003). By the composition of Web services in SOA-driven business Webs, value is co-created by co-opetitively acting providers.

These business Webs exhibit two kinds of actors: shapers and adapters (Hagel 1996), where shapers antedate the upcoming developments and drive changes in current situations, e.g., by actively taking part in definition of standards, and adapting pricing schemes. In the contrast, adapters follow changes and are regarded to be price-takers.

[1] http://www.salesforce.com/.

Examples for economic considerations of providers participating in the market include approaches for the efficient allocation of service invocations (Blau 2009) that relies on algorithmic mechanism design (cf. Nisan and Ronen (2001)). The allocation is executed by means of a mechanism, which is a special form of a path auction (cf. Archer and Tardos (2007); Hershberger and Suri (2001); Immorlica et al. (2005)) that is applied to an SVN. However, the mechanism introduced in Blau (2009) allows for service failure as some exogenously given variable, whereas the work at hand considers service delivery and the respective effort as an endogenous factor.

The cooperation of service providers for the integration of service modules is regulated by means of SLAs, which comprise the concerned parties, a description of the respective service on a functional as well as a non-functional and quality oriented level. Furthermore, the price for service delivery as well as a penalty that is applied in case of SLA violation are defined. The specification of the latter ones is under constant discussion and many approaches can be observed in practice. Becker et al. (2008) describe a methodology which sets incentives for providers to adhere to the agreed levels of quality. Penalties are computed by a k-pricing scheme, which is part of an incentive-compatible mechanism, which can only be applied in a setting of a multitude of independent services but not for complex services that are built from service modules. Macias and Guitart (2008) introduce a different approach that is based on reputation, analogously to the ones implemented in online market places like ebay.[2] These decentral environments require to judge providers based on their performance in past agreements and to aggregate this knowledge into a term that states the provider's reputation for future cooperations.

In Filipova-Neumann et al. (2010), the model of this chapter is applied to a scenario in which a central integrator that acts as an intermediary between a consumer and multiple service providers. The intermediary purchases complementary services from the providers to compose a complex service and sets the prices and penalties for all parties on both sides of the market. Since single service providers along the value chain in this chapter establish an SLA with a buyer and a seller of a service, each one of them can also be regarded as an intermediary. A very general overview of the economic literature on intermediation theory is given in Spulber (1999).

By setting the price and penalty for the service at the preceding value-added step, service providers are incentivized to exert the desired level of effort. In this respect, the model in this chapter is related to the vast literature on principal–agent models within the theory of the firm, which examines delegation and incentives outside and within the boundaries of organizations. An introduction to principal–agent literature may be found e.g., in Ross (1973), Stiglitz (1975), Shavell (1979), and Holmström (1979). Principal–agent relationships are characterized by one participant (the principal), which employs another participant (the agent) to fulfill tasks for the former or on their behalf. These structures allow to investigate on setting incentives and distributing risk on the agent while taking imperfect

[2] www.ebay.com.

information on effort and its impact on SLAs into account. The interest of the model in this chapter also lies in the optimal structure of agreements between the service providers that can be regarded alternatively as the principal or the agent depending on whether they appear as the buyer or the provision of the service in the agreement at hand. The subjects under consideration in this chapter are regarded to be risk neutral, that is, they exhibit utility functions that are linear in profit. Therefore, the transfer of risk is not taken into account. In that case, the purpose of designing the SLAs boils down to setting the proper incentives. In contrast to the above articles, this chapter focuses on the interaction between numerous providers along a value chain and how it influences the design of SLAs.

6.3 Model and Assumptions

The assumptions of the model are as follows. There are n value-added steps of service provision denoted by $i = 1, 2, \ldots, n$. At each value-added step, a single service s_i is provided which is combined with the services from the preceding value-added steps to compose the complex service $s_c = (s_1, \ldots, s_i, \ldots, s_n)$. There is perfect competition among service providers at each step of the value chain. That is, services within a value-added step are regared as homogeneous, while they are complementary across the value-added steps. A service provider at value-added step i is denoted by A_i. Hence, a service provider A_i who himself provides exclusively service s_i establishes an SLA with the upstream service provider A_{i+1} in order to purchase the service $s_{c_{i+1}} = (s_{i+1}, s_{i+2}, \ldots, s_n)$ which he integrates with the service s_i. He also enters into an agreement with a downstream provider A_{i-1} for the sale of service $s_{c_i} = (s_i, s_{i+1}, \ldots, s_n)$. The utmost downstream service provider A_1 establishes an agreement directly with the customer for the sale of the complex service $s_c = (s_1, \ldots, s_i, \ldots, s_n)$. The value chain is depicted in Fig. 6.1.

Further, it is assumed that a service s_i can be either delivered in the quality agreed in the service level agreement or the service provision fails. Hence, service failure translates directly into a violation of the established SLA and consequently, into due penalties. Service failure is uncertain and depends on a provider's effort, where $e_i \in [\underline{e_i}, \bar{e_i}]$ denotes the effort of service provider A_i (e.g., personnel costs or costs of investing in new hardware). The probability of service delivery of

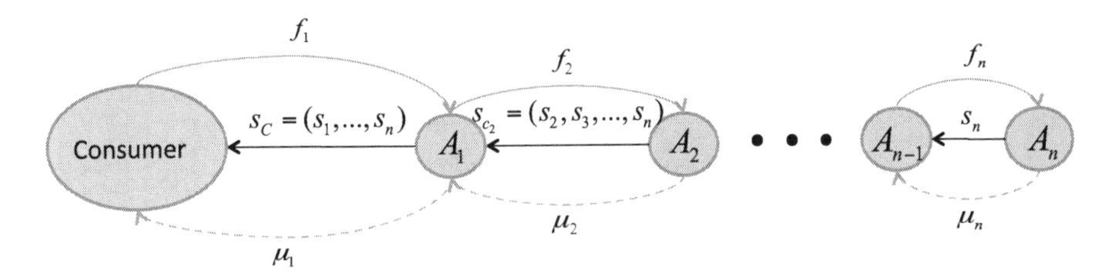

Fig. 6.1 Service level agreements between service providers and a consumer

service s_i, $p_i(e_i) \in (0, 1)$, is assumed to be increasing and concave in effort, $p_i'(e_i) > 0$, $p_i''(e_i) < 0$. Accordingly, the probability of failure of service s_i, which is decreasing in the provider's effort, is $1 - p_i(e_i)$.

It is assumed that services are perfect complements. Thus, the consumer derives utility of the complex service, only if all single services are provided. The consumer's utility is v if the composite service $s_C = (s_1, \ldots, s_i, \ldots, s_n)$ is delivered and zero otherwise, i.e., if any of the services s_i fails. The variable v is assumed to be common knowledge.

The required steps before single services can be integrated into a complex service s_C which is ready for consumption by the customer, are the establishment of all SLAs between the service providers and the customer, the choice and exertion of effort levels by the service providers and the delivery of the single services s_i.

A crucial assumption of the SLAs is that the price of service provision and the penalty in case of service failure are binding and enforceable. Let f_i denote the price, which service provider A_i receives from service provider A_{i-1} for the provision of services $(s_i, s_{i+1}, \ldots, s_n)$. At the first value-added step f_1 is the price paid by the consumer to service provider A_1 for the provision of the complex service s_C. Denote by μ_i the penalty which service provider A_i pays to service provider A_{i-1} in case that he fails to provide $(s_i, s_{i+1}, \ldots, s_n)$. μ_1 is the penalty which the service provider A_1 pays to the consumer in case of failure of the complex service s_C.

Without loss of generality, there are no costs of service provision except for the costs of effort. These are measured per unit of effort, are common knowledge, and are denoted by c_i, where i is the index for the service provider.

The expected payoffs from participating in the SLAs as depicted in Fig. 6.1 and social welfare (the total surplus) can be specified as follows.

Definition 6.1 (Expected payoffs). Denote by

$$E(\pi_i) = f_i - f_{i+1} - c_i \cdot e_i - \mu_i \cdot \left(1 - \prod_{j=i}^{n} p_j(e_j)\right) + \mu_{i+1} \cdot \left(1 - \prod_{j=i+1}^{n} p_j(e_j)\right), \tag{6.1}$$

where $(f_{i+1}, \mu_{i+1}) = (0, 0)$ for $i = n$, since there is no upstream contracting party for service provider A_n,

$$E(u) = v \cdot \prod_{i=1}^{n} p_i(e_i) - f_1 + \left[1 - \prod_{i=1}^{n} p_i(e_i)\right] \mu_1 \tag{6.2}$$

and by

$$\begin{aligned} W &= \sum_i E(\pi_i) + E(u) \\ &= v \cdot \prod_i p_i(e_i) - \sum_i c_i \cdot e_i \end{aligned} \tag{6.3}$$

the expected profit $E(\pi_i)$ of service provider A_i, the expected utility $E(u)$ of the consumer, and social welfare W, respectively.

The expected profit of the service provider $E(\pi_i)$, (6.1), consists of the following components. The first term $f_i - f_{i+1}$ corresponds to the difference between the price f_i the provider receives from service provider A_{i-1} for the provision of $s_{C_i} = (s_i, s_{i+1} \dots s_n)$ and the price he pays to service provider A_{i+1} for the provision of $s_{C_{i+1}} = (s_{i+1}, s_{i+2} \dots s_n)$. The second term $c_i \cdot e_i$ denotes the costs incurred by the provider for exerting effort on service s_i, and the last term $-\mu_i \cdot \left(1 - \prod_{j=i}^{n} p_j(e_j)\right) + \mu_{i+1} \cdot \left(1 - \prod_{j=i+1}^{n} p_j(e_j)\right)$ is the expected net penalty in case of service failure. The provider pays an expected penalty of $\mu_i \cdot \left(1 - \prod_{j=i}^{n} p_j(e_j)\right)$ for failure of the complex service s_{C_i}, i.e., of any one of the services $s_i, s_{i+1}, \dots s_n$, but he also receives the expected payment $\mu_{i+1} \cdot \left(1 - \prod_{j=i+1}^{n} p_j(e_j)\right)$ for failure of the complex service $s_{C_{i+1}}$, i.e., of any one of the services $s_{i+1}, s_{i+2} \dots s_n$. Note that the single service failure are statistically independent. Hence, the probability of failure of any one of the services $s_i, s_{i+1}, \dots s_n$ is $\left(1 - \prod_{j=i}^{n} p_j(e_j)\right)$. The consumer's utility $E(u)$ (see (6.2)) consists of their valuation for the complex service, which they obtain with probability $\prod_i p_i(e_i)$, less the net payments to the service provider A_1: the price paid f_1 net of the expected penalty received $[1 - \prod_i p_i(e_i)]\mu_1$. Finally, social welfare (6.3) is defined as the sum of all parties' payoffs. Note that the only endogenous variable left in the objective function is e_i, $i = 1, 2, \dots, n$ since prices and penalties are only redistributed among agents without altering total welfare.

In the scenario considered in this chapter, there is a consumer who demands a complex service from service providers who offer perfectly complementary single services across the value-added steps and who compete perfectly for the provision of a homogeneous single service at each value-added step. In Fig. 6.1, A_i, where $i = 1, 2, \dots, n$, represents a provider who is arbitrarily chosen among the numerous providers competing at the corresponding value-added step i to contribute to the complex service. As a result, it is the consumer who possesses all bargaining power in the value chain and service providers at each upstream step behave as price takers with respect to the purchasers at the downstream step. This is easily seen for the consumer and the utmost downstream service provider A_1, but the pattern continues along the value chain up to the utmost upstream service provider A_n.

The sequence of moves is summarized as follows:

Stage 1. The consumer requests the complex service s_C and sets the price and penalty (f_1, μ_1)

Stage 2. Provider A_1 sets (f_2, μ_2)

...

Stage n. Provider A_{n-1} sets (f_n, μ_n)

Stage $n + 1$. Provider A_n either accepts or rejects (f_n, μ_n)

Stage $n + 2$. Provider A_{n-1} either accepts or rejects (f_{n-1}, μ_{n-1})

...

Stage $2n$. Provider A_1 either accepts or rejects (f_1, μ_1)

Stage $2n + 1$. Service providers $A_1, A_2, \ldots, A_n$ choose $e_1, e_2, \ldots, e_n$, respectively

Stage $2n + 2$. Services are delivered or fail and the agreed upon payments are realized.

Prices, and especially penalties, serve as a tool to incentivize upstream providers to exert the required levels of effort. Even though the level of effort is not part of the service level agreement, by varying the penalties, the downstream provider is able to induce the upstream provider to exert more or less effort. This in turn has an immediate effect on the probability of failure both of the single services as well as on the complex service, and the prices payed between the parties of the SLAs. In order to find the optimal prices and penalties, the above sequence of moves is solved with backward induction in the following sections.

6.4 Choosing the Levels of Effort

In stage $2n + 1$ from the previous section, service providers decide how much effort to exert once they have entered the agreements with a downstream service provider (or the consumer, respectively) and an upstream service provider (if such exists) for given prices and penalties f_i, f_{i+1}, and μ_i, μ_{i+1}. Each service provider A_i will choose the effort level e_i, which maximizes the expected profit equation (6.1). The first order condition is

$$p_i{}'(e_i) \cdot \mu_i \cdot \prod_{j=i+1}^{n} p_j(e_j) = c_i. \tag{6.4}$$

The left-hand side of this equation is the marginal benefit of effort, the decline of expected penatly. The right-hand side is the marginal cost of effort.

Denoting by g_i the inverse function of $p_i'(e_i)$ one can transform the above equation into an explicit solution for the effort level. Thus, (6.4) is equivalent to

$$e_i = g_i\left(\frac{c_i}{\mu_i \cdot \prod_{j=i+1}^{n} p_j(e_j)}\right). \tag{6.5}$$

Note that the marginal benefit of effort depends on the upstream probability of service failure. This is because provider A_i will be obliged to pay the penalty if his own or any one of the single services of an upstream service provider fails. Holding other things constant, a provider's effort e_i depends positively on the effort levels of upstream service providers, $e_{i+1}, e_{i+2}, \ldots, e_n$, and negatively on the number of upstream service providers. To obtain all levels of effort, for given prices and penalties, one needs to solve the system of all n equations (6.5).

In contrast, the n socially optimal levels of effort, which maximize total surplus are found from the n first-order conditions to (6.3). The first order conditions are

$$p_i'(e_i)v \prod_{j \neq i} p_j(e_j) = c_i,\ i, j = 1, \ldots, n, \quad i \neq j, \tag{6.6}$$

Denote the levels of effort, which solve the above system of n equations, by e_i^*, $i = 1, 2, \ldots, n$. As before, they equate the marginal benefit to the marginal cost of effort. However, the marginal benefit from a social perspective takes account of all other service providers and therefore the term includes the the effort of all remaining service providers A_j, $j \neq i$. The difference between the private choice of effort (6.4) and the socially optimal choice of effort, (6.6), shows that the service providers' behavior causes externalities, which providers do not take into account. In particular, they do not consider the complementarity of services and the fact that all services need to be provided for the consumer to derive utility. This result is summarized below:

Proposition 6.1. *The service providers' choice of effort causes external effects and the effort levels that service providers choose differ, in general for given prices and penalties, from the socially optimal levels.*

Proof. As argued above, this result immediately follows from the comparison of the first order conditions (6.4) and (6.6).

As before, the inverse function $g(\cdot)$ allows for phrasing the explicit solutions to the socially optimal effort levels

$$e_i = g_i \left(\frac{c_i}{v \prod_{j \neq i} p_j(e_j)} \right). \tag{6.7}$$

After having determined the optimal levels of effort for private as well as socially optimal choice, respective prices an penalties are derived in the following section.

6.5 Choosing Prices and Penalties

In the previous section, it was shown that the private choice of effort of service providers and what would be optimal from a social perspective will not necessarily coincide for given prices and penalties. If the penalty is too low, $\mu_i < v \prod_{j \neq i} p_j(e_j^*)/(\prod_{j=i+1}^{n} p_j(e_j))$, provider $A_i's$ effort will be insufficient from a social perspective. The probability of service delivery would be too low in this case. If the penalty is too high, $\mu_i > v \prod_{j \neq i} p_j(e_j^*)/(\prod_{j=i+1}^{n} p_j(e_j))$, the provider $A_i's$ effort will also be too high compared to the level which would be optimal for welfare maximization. Characterizing effort as "too high" makes sense when realizing that it comes at a cost for service providers which is included in (6.3). Hence, the socially optimal levels of effort will be the result of a trade-off between the higher probability of service delivery and the higher costs of effort.

The socially optimal penalty μ_i^*, which aligns the private incentives of service providers with the socially optimal solution for the effort levels, is such that equates the private (6.4) and social (6.6) marginal benefist of effort. This is equivalent to

$$\mu_i^* = \frac{v \cdot \prod_{j=1,,j\neq i}^{n} p_j(e_j^*)}{\prod_{j=i+1}^{n} p_j(e_j^*)} = v \cdot \prod_{j=1}^{i-1} p_j(e_j^*), \tag{6.8}$$

where e_j^* are the socially optimal effort levels as calculated in the previous section (see (6.7)). Directly following from (6.8) one can state

Proposition 6.2. *The socially optimal penalty μ_i^* depends negatively on the number of downstream providers and, ceteris paribus, positively on the downstream providers' socially optimal probability of service delivery.*

In other words, the proposition states that from a welfare perspective, the penalty a provider A_i agrees to pay in case of failure of $(s_i, s_2, \ldots s_n)$ to his neighbouring downstream provider A_{i-1} should be higher than the penalty that he contracts to receive from his neighbouring upstream provider A_{i+1} for failure of (s_{i+1}, s_{i+2}, s_n), $\mu_i^* > \mu_{i+1}^*$. Moreover, this result holds even for the penalties in expectation, i.e., when considering the corresponding probabilities of failure. The expected penalty service provider A_i pays is higher than the expected penalty that he receives,

$$\mu_i^* \cdot \left(1 - \prod_{j=i}^{n} p_j(e_j^*)\right) > \mu_{i+1}^* \cdot \left(1 - \prod_{j=i+1}^{n} p_j(e_j^*)\right)$$

because $\left(1 - \prod_{j=i}^{n} p_j(e_j^*) > 1 - \prod_{j=i+1}^{n} p_j(e_j^*)\right)$.

To calculate the prices and penalties resulting from the interaction of the service providers, consider stage $2n$ from Sect. 6.3. Provider A_1 will accept the consumer's offer of (μ_1, f_1), for given price and penalty quotes from the previous stages, if thereby $E(\pi_1 \geq 0)$, where 0 is the outside option of the service provider considering that there is perfect competition among service providers at the same value-added step. Continuing the sequence of moves in the same fashion along the value chain, at stage $n+1$ provider A_n accepts provider A_{n-1}'s offer (μ_n, f_n) if thereby $E(\pi_n) \geq 0$. At stage n provider A_{n-1} sets (μ_n, f_n) such that he maximizes his profit

$$max_{\mu_n, f_n} E(\pi_{n-1}) \tag{6.9}$$

subject to

$$E(\pi_n) \geq 0 \tag{6.10}$$

$$E(\pi_{n-1}) \geq 0 \tag{6.11}$$

$$\ldots \tag{6.12}$$

$$E(\pi_1) \geq 0 \tag{6.13}$$

$$e_i = argmax\, E(\pi_i), i = 1, 2, \ldots, n \tag{6.14}$$

This is equivalent to

$$max_{\mu_n, f_n} E(\pi_{n-1}) = f_{n-1} - f_n - c_{n-1} \cdot e_{n-1} \\ - \mu_{n-1} \cdot (1 - p_{n-1}(e_{n-1}) \cdot p_n(e_n)) + \mu_n \cdot (1 - p_n(e_n)) \quad (6.15)$$

subject to

$$E(\pi_n) = f_n - c_n \cdot e_n - \mu_n \cdot (1 - p_n(e_n)) \geq 0 \quad (6.16)$$

$$e_i = argmax\, E(\pi_i), i = 1, 2, \ldots, n \quad (6.17)$$

because only the zero-profit constraint for the neighbouring upstream provider A_n is binding. Substituting from (6.5) and solving, one obtains $\mu_n = \mu_{n-1} \cdot p_{n-1}(e_{n-1})$ and $f_n = c_n \cdot e_n + \mu_n \cdot [1 - p_n(e_n)]$. Moving backwards from stage to stage, at stage 1 the consumer's maximization problem is

$$max_{\mu_1, f_1} E(u) = v \cdot \prod_{i=1}^{n} p_i(e_i) - f_1 + \left[1 - \prod_{i=1}^{n} p_i(e_i)\right] \mu_1 \quad (6.18)$$

subject to

$$E(\pi_1) = f_2 - f_1 - c_1 \cdot e_1 - \mu_1 \cdot \left(1 - \prod_{j=i}^{n} p_j(e_j)\right) + \mu_2 \cdot \left(1 - \prod_{j=2}^{n} p_j(e_j)\right) \\ \geq 0 \quad (6.19)$$

$$\mu_i = \mu_{i-1} \cdot p_{i-1}(e_{i-1}), i = 2, 3, \ldots, n \quad (6.20)$$

$$f_i = \sum_{i}^{n} c_i \cdot e_i + \mu_i [1 - \prod_{j=i}^{n} p_j(e_j)], i = 2, 3, \ldots, n \quad (6.21)$$

$$e_i = argmax\, E(\pi_i), i = 1, 2, \ldots, n \quad (6.22)$$

One obtains that $\mu_1 = v$ and $f_1 = \sum_{i=1}^{n} c_i \cdot e_i + v[1 - \prod_{i=1}^{n} p_i(e_i)]$. One can easily verify that the thus obtained penalties and effort levels, respectively, correspond to the socially optimal solutions, $\mu_i = \frac{v \cdot \prod_{j=1, j \neq i}^{n} p_j(e_j^*)}{\prod_{j=i+1}^{n} p_j(e_j^*)} = v \cdot \prod_{j=1}^{i-1} p_j(e_j^*) = \mu_i^*$. This is no surprise given perfect competition at each stage and risk neutrality of agents.

6.6 Exogenous Impact on Effort

In the previous sections, the equilibrium prices, penalties, and levels of effort were derived, which were shown to coincide with the socially optimal values. In this section, comparative statics analysis is performed in order to examine how those endogenously determined values are affected by the factors which are exogenous to the model.

6.6.1 Comparative Statics Analysis with Two Service Providers

For tractability, the further analysis is performed for the case of only two service providers, $n = 2$. This situation is depicted in Fig. 6.2.

The optimality conditions for the effort level, (6.7), reduce to only two equations, $e_i(e_j)$,

$$e_i = g_i\left(\frac{c_i}{v p_j(e_j)}\right), \tag{6.23}$$

where $i, j = 1, 2, i \neq j$.

The interdependence (outside the equilibrium) between the effort levels of the service providers is expressed in the "reaction function." The effort level of one service provider as a function of the effort level of the other service provider, $e_i(e_j)$ can be determined by differentiating equation (6.23) with respect to e_j

$$\frac{de_i}{de_j} = \frac{dg_i}{de_j} = g_i{}'(\cdot)\frac{c_i}{v}(-1)[p_j(e_j)]^{-2} p_j{}'(e_j) \tag{6.24}$$

$$= -\frac{1}{p_i{}''(e_i)}\frac{c_i}{v}[p_j(e_j)]^{-2} \cdot p_j{}'(e_j) > 0. \tag{6.25}$$

This interdependence shows that the effort levels behave complementary to each other. Service provider A_i should increase his effort level in response to an increase of service provider A_j's effort level and vice versa. The reason for this complementarity is related to the assumption of complementary services. An increase of e_j also increases the probability of service delivery $p_j(e_j)$. This, in turn, raises ceteris paribus the probability that the complex service will be deliverd and, hence, the marginal (social) benefit of effort for provider A_i in (6.6). Because marginal benefit is decreasing in effort, for the optimality condition to hold, the effort level e_i must rise.

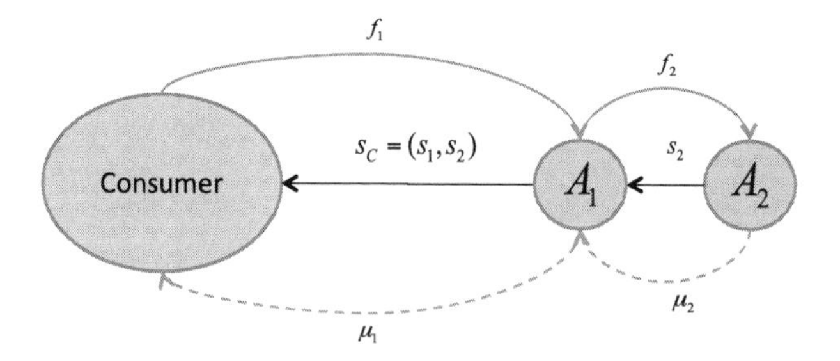

Fig. 6.2 Service level agreements between service providers and a consumer, $n = 2$

This result expresses the existence of positive external effects which service providers cause to each other by exerting effort.

For the comparative statics analysis, the implicit function theorem will be applied. To this purpose, rewrite the first-order conditions, (6.23), to define the implicit functions

$$F_i : -e_i + g_i\left(\frac{c_i}{vp_j(e_j)}\right) = 0 \tag{6.26}$$

where $i, j = 1, 2, i \neq j$.

Further, denote by

$$H = \left(\frac{\partial^2 W}{\partial e_i \partial e_j}\right), i, j = 1, 2, \tag{6.27}$$

$$= \begin{pmatrix} \frac{\partial F_1}{\partial e_1} & \frac{\partial F_1}{\partial e_2} \\ \frac{\partial F_2}{\partial e_1} & \frac{\partial F_2}{\partial e_2} \end{pmatrix} \tag{6.28}$$

the Hesse matrix of the welfare maximization, where the latter equivalence results from (6.26) corresponding to the first order conditions, $\frac{\partial W}{\partial e_i} = 0$. The second-order sufficient conditions of the service integrator's problem require the Hesse matrix to be negative definite, $detH_1 = \frac{\partial^2 E(\pi_i)}{\partial e_i^2} < 0$, $detH > 0$. The former condition is always satisfied, since (see (6.26))

$$\frac{\partial^2 E(\pi_i)}{\partial e_i^2} = \frac{\partial F_i}{\partial e_i} = -1 < 0 \tag{6.29}$$

while the latter condition is satisfied if

$$detH = \frac{\partial F_1}{\partial e_1}\frac{\partial F_2}{\partial e_2} - \frac{\partial F_1}{\partial e_2}\frac{\partial F_2}{\partial e_1} = 1 - \frac{de_1}{de_2}\frac{de_2}{de_1} > 0. \tag{6.30}$$

For the further analysis assume that this condition is satisfied.

The results of the comparative statics analysis are summarized in the following propositions.

The first proposition introduces the impact of a change in marginal costs on optimal effort levels.

Proposition 6.3. *The higher are the marginal costs c_i of service provider A_i, the lower should be the optimal levels of effort of both service provider A_i and service provider A_j,*

$$\frac{\partial e_i{}^*}{\partial c_i} < 0, \frac{\partial e_j{}^*}{\partial c_i} < 0 \quad i, j = 1, 2, i \neq j, \tag{6.31}$$

and, the lower should be the optimal penalty of the upstream service provider,

$$\frac{\partial \mu_2{}^*}{\partial c_2} < 0, \frac{\partial \mu_2{}^*}{\partial c_1} < 0. \tag{6.32}$$

Proof. By the implicit function theorem,

$$\frac{\partial e_i^*}{\partial c_i} = \frac{det\, J_{e_i,c_i}}{det\, H}, \tag{6.33}$$

where

$$J_{e_i,c_i} = \begin{pmatrix} -\frac{\partial F_i}{\partial c_i} & \frac{\partial F_i}{\partial e_j} \\ -\frac{\partial F_j}{\partial c_i} & \frac{\partial F_j}{\partial e_j} \end{pmatrix}. \tag{6.34}$$

Taking the respective derivatives of (6.26) for i, j, one obtains

$$\frac{\partial F_i}{\partial c_i} = g_i'(\bullet) \cdot \frac{1}{v \cdot p_j(e_j)} < 0, \tag{6.35}$$

$$\frac{\partial F_j}{\partial c_i} = 0 \tag{6.36}$$

$$\frac{\partial F_i}{\partial e_j} = \frac{de_i}{de_j} > 0, \tag{6.37}$$

which is the expression in (6.24), and

$$\frac{\partial F_j}{\partial e_j} = -1 < 0. \tag{6.38}$$

Thus,

$$det\, J_{e_i,c_i} = \left(-\frac{\partial F_i}{\partial c_i}\right) \cdot \frac{\partial F_j}{\partial e_j} - \frac{\partial F_i}{\partial e_j} \cdot \left(-\frac{\partial F_j}{\partial c_i}\right) \tag{6.39}$$

$$= g_i'(\cdot)\frac{1}{v \cdot p_j(e_j)} < 0 \tag{6.40}$$

and, hence, $\frac{de_i^*}{dc_i} < 0$. Similarly, $\frac{\partial e_j^*}{\partial c_i} = \frac{det\, J_{e_j,c_i}}{det\, H}$, where

$$J_{e_j,c_i} = \begin{pmatrix} \frac{\partial F_i}{\partial e_i} & -\frac{\partial F_i}{\partial c_i} \\ \frac{\partial F_j}{\partial e_i} & -\frac{\partial F_j}{\partial c_i} \end{pmatrix}. \tag{6.41}$$

For the derivatives one obtains the expressions

$$\frac{\partial F_i}{\partial e_i} = -1 \tag{6.42}$$

$$\frac{\partial F_j}{\partial e_i} = \frac{de_j}{de_i} > 0, \tag{6.43}$$

which is the expression in (6.24) with reversed index. Thus,

$$det\, J_{e_j,c_i} = -1 \cdot 0 - \left(-g_i'(\cdot) \cdot \frac{1}{v \cdot p_j(e_j)}\right) \cdot \frac{de_j}{de_i} \quad (6.44)$$

$$= g_i'(\cdot) \cdot \frac{1}{v \cdot p_j(e_j)} \cdot \frac{de_j}{de_i} < 0. \quad (6.45)$$

Thus, $\frac{\partial e_j^*}{\partial c_i} < 0$. Obviously, these results hold for the other service provider too, such that $\frac{\partial e_j^*}{\partial c_j} < 0$ and $\frac{\partial e_i^*}{\partial c_j} < 0$. From the expressions for the optimal penalties, (6.8), one obtains

$$\frac{d\mu_2^*}{de_1^*} = v \cdot p_1'(e_1^*) > 0. \quad (6.46)$$

Hence, using the above results, it follows that

$$\frac{\partial \mu_2^*}{\partial c_2} = \frac{d\mu_2^*}{de_1^*} \cdot \frac{\partial e_1^*}{\partial c_2} < 0 \quad (6.47)$$

and

$$\frac{\partial \mu_2^*}{\partial c_1} = \frac{d\mu_2^*}{de_1^*} \cdot \frac{\partial e_1^*}{\partial c_1} < 0. \quad (6.48)$$

While the penalty of the downstream provider is independent of the costs of effort, the penalty of the upstream provider is positively related to both the costs of the upstream as well as the downstream provider. For the relative size of the sensitivity of both providers' optimal effort levels to provider A_i's costs c_i, one obtains

Proposition 6.4. *The effect of provider A_i's marginal costs of effort c_i is stronger on provider A_i's own optimal level of effort e_i^* than on the the other provider A_j's optimal level of effort e_j^*,*

$$\frac{\partial e_i{}^*}{\partial c_i} < \frac{\partial e_j{}^*}{\partial c_i} < 0, \quad (6.49)$$

if

$$\frac{de_j}{de_i} < 1 \quad (6.50)$$

holds and vice versa.

Proof. From Proposition 6.3's results it follows that $\frac{\partial e_i{}^*}{\partial c_i} < \frac{\partial e_j{}^*}{\partial c_i} < 0$ is equivalent to $det\, J_{e_i,c_i} < det\, J_{e_j,c_i}$. As can be seen in the respective expressions in (6.39) and (6.44), this is equivalent to $\frac{de_j}{de_i} < 1$.

For the sensitivity of provider A_2's optimal penalty to provider A_i's, $i = 1, 2$, marginal costs one obtains for symmetric providers with identical distributions of service delivery $p_1(e) = p_2(e)$

Proposition 6.5. *If* $\frac{de_j}{de_i} < 1,\ i, j = 1, 2,\ i \neq j$ *the providers are symmetric,* $p_1(e) = p_2(e)\,\forall e$ *and initially* $e_1^* = e_2^*$*, then provider* A_2*'s optimal penalty* μ_2^* *is more sensitive to provider* A_1*'s marginal costs of effort* c_1 *than to* A_2*'s marginal costs of effort* c_2

$$\frac{\partial \mu_2^*}{\partial c_1} < \frac{\partial \mu_2^*}{\partial c_2} < 0. \tag{6.51}$$

Proof. By proposition 6.4 $\frac{\partial e_2{}^*}{\partial c_2} < \frac{\partial e_1{}^*}{\partial c_2} < 0$ and by symmetry of service providers, $\frac{\partial e_2{}^*}{\partial c_2} = \frac{\partial e_1{}^*}{\partial c_1}$. Hence, $\frac{\partial e_1{}^*}{\partial c_1} < \frac{\partial e_1{}^*}{\partial c_2} < 0$. For the penalties, one obtains that $\frac{\partial \mu_2^*}{\partial c_1} < \frac{\partial \mu_2^*}{\partial c_2} < 0$.

For the effect of the consumer valuation on providers' optimal effort levels and penalties one obtains

Proposition 6.6. *The higher is the consumer's valuation v for the complex service, the higher should be the optimal levels of effort and the optimal penalties of both providers.*

Proof. The effect of the consumer's valuation on the effort levels is $\frac{de_i^*}{dv} = \frac{det\, J_{e_i,v}}{det\, H}$, where

$$J_{e_i,v} = \begin{pmatrix} -\frac{\partial F_i}{\partial v} & \frac{\partial F_i}{\partial e_j} \\ -\frac{\partial F_j}{\partial v} & \frac{\partial F_j}{\partial e_j} \end{pmatrix} \tag{6.52}$$

$$\frac{\partial F_i}{\partial v} = g_i'(\cdot)\frac{c_i}{p_j(e_j)} \cdot (-1) \cdot v^{-2} > 0. \tag{6.53}$$

The same expression with reverse index holds for $\frac{\partial F_j}{\partial v}$. Thus,

$$J_{e_i,v} = \left(-g_i'(\cdot)\frac{c_i}{p_j(e_j)} \cdot (-1) \cdot v^{-2}\right) \cdot (-1) \tag{6.54}$$

$$-\left(\frac{de_i}{de_j}\right) \cdot \left(-g_j'(\cdot)\frac{c_j}{p_i(e_i)} \cdot (-1) \cdot v^{-2}\right) > 0 \tag{6.55}$$

and, hence, $\frac{\partial e_i^*}{\partial v} > 0$. By symmetry, this holds for $\frac{\partial e_j^*}{\partial v} > 0$. For the penalties one obtains $\frac{\partial \mu_1^*}{\partial v} = 1$ and

$$\frac{\partial \mu_2^*}{\partial v} = p_1(e_1^*) + v \cdot p_1'(e_1^*) \cdot \frac{\partial e_1^*}{\partial v} > 0. \tag{6.56}$$

Consider an exogenous increase of the marginal costs of effort c_1 of the downstream service provider. With $\frac{de_2}{de_1} > 1$, the sensitivity of the upstream service provider A_2's optimal effort e_2^* to the downstream provider A_1's optimal effort e_1^* is relatively weak. As can be seen from (6.24), this sensitivity depends on all variables of the model. Therefore, the external effect of provider A_1's choice of effort on the marginal social benefit of provider A_2's effort is relatively weak. As a result, the exogenous shock on the costs c_1 will affect the upstream service provider A_2 only weakly. The reasoning is reversed, if the externality between the service providers' effort levels is strong enough, $\frac{de_2}{de_1} > 1$. In that case, an exogenous shock causing c_1 to increase will more strongly affect the upstream service provider A_2 rather than the downstream service provider A_1 who has suffered the cost increase.

6.6.2 *Increasing the Number of Providers*

Even though no comparative statics analysis will be performed for n service providers, the effect of increasing the number of service providers on the optimal levels of effort as well as on the resulting penalties should be examined.

Consider the socially optimal levels of effort e_i^*, $i = 1, \ldots, n$ for which the necessary conditions (6.6) $p_i'(e_i{}^*) \cdot \prod_{j \neq i} p_j(e_j{}^*) \cdot v = c_i$ hold. Adding one service provider $n + 1$ while holding everything else constant will reduce the probability of delivery of the complex service and, hence, the marginal benefit of effort of service provider A_i on the left-hand side of the first order condition, $p_i'(e_i{}^*) \cdot p_{n+1}(e_{n+1}) \cdot \prod_{j \neq i} p_j(e_j{}^*) \cdot v < c_i$. This occurs even if $p_{n+1}(e_{n+1})$ is close to one. As a result, the optimal level of effort $e_i{}^*$ needs to be decreased. The effort levels exerted by each single service provider as well as the optimal penalties decrease with the number of single services involved in the composition of the complex service. The direct consequence is that the probability of service delivery of any single service, and even more of the complex service, suffers with an increasing number of complementary services.

6.7 Discussion and Conclusion

The setting in this chapter models service provision with service providers establishing SLAs successively along the value chain of a complex service. The consumer is assumed to have the full bargaining power while service providers compete perfectly at each value-added step for the provision of a homogeneous single service. Service delivery depends on external (random) factors and on the effort of service providers. A higher level of effort increases the probability of delivery of the single service. It is also assumed that single services are perfect complements to each other implying that the consumer derives utility, only if all single services constituting a complex service are delivered. The complementarities between single services across the value-added steps are the reason for the existence of positive externalities when providers choose their effort levels. A higher effort of one service provider also increases the probability of delivery of the complex service. Penalties, which upstream service providers pay for service failure to downstream service providers or to the consumer, as the case might be, serve as as an incentive tool along the value chain. By means of penalties, combined with prices accordingly, the consumer and downstream service providers align the incentives of upstream service providers with their own objectives.

It is shown that, in the setting of the model, the equilibrium allocation of effort and payments maximize the system's welfare. In particular, downstream service providers will not seek to maximize the upstream service providers' effort and the probability of upstream service delivery. The optimal levels of effort, which coincide from a private and social perspective, are rather the result of a trade-off between benefits and costs of effort.

The comparative statics analysis shows that an increase of one service provider's effort level causes the marginal social benefit of effort of the remaining service providers to increase. The resulting penalty for service failure of a particular service provider is related to the number of downstream providers along the value chain and to the probability of service delivery of the downstream providers. The penalty of a given provider along the value chain increases with the probability of delivery of the downstream single services, ceteris paribus, and decreases with the number of downstream value-added steps. That is, the further downstream a provider is located along the value chain, the less sensitive is the penalty that he should pay to the downstream recipient provider or to the consumer (and also the penalty he will receive in case of service failure of the upstream provider) to the number of single services in the complex service and to the probability of service delivery of the single services.

The complementarity of services is the reason for a negative relationship between the number of single services involved in the composition of the complex service and the levels of effort. The higher is the number of single services, the lower will be the equilibrium effort levels and probabilities of delivery of the single services and, hence, of the complex service. This result by no account should imply that the number of single services should be held small. It only shows that a rising number of services is accompanied by rising complexity and vulnerability to failure. In addition, the importance of any single provider decreases with the total number of providers, which is the reason for a given provider to exert less effort. Even though this result can be characterized as a form of free riding, as was argued before, the social perspective does not impose higher effort levels than what the providers choose in equilibrium from their private perspective. With increasing number of single services, the penalties for service failure will decrease for all but the utmost downstream service providers.

The comparative statics analysis yields some results as to the sensitivity of the effort levels and penalties to the exogenous variables of the model. It is shown that the costs of a single service provider influence all providers' effort levels and the penalties of upstream service providers. Increasing costs of one provider lead, ceteris paribus, to a decrease of the optimal effort levels of all providers, vice versa for the case of decreasing costs. Increasing costs across different providers might result from exogenous factors such as rising electricity tariffs, for example, and would make service providers decrease all penalties but the utmost downstream one. In contrast, the declining hardware costs will generally increase the equilibrium penalties and the probabilities of service delivery.

While the *direction* of this effect is always the same for all service providers, its *size* differs in general and depends on the strength of the externalities. If costs of any provider increase, then all providers will decrease their levels of effort. If the external effects between service providers are weak enough ($\frac{de_j}{de_i} < 1$), then the sensitivity of a provider's effort e_i to a change in his own costs c_i is stronger than the sensitivity of another provider's effort level e_j to this change.

An unambiguous comparison of the size of the effects on penalties can be performed only for the case of symmetric service providers. In the case of two service providers, an exogenous increase of the costs of the downstream service provider causes the penalty of the upstream service providers to decrease more strongly than when his own costs increase.

The impact of the customer's valuation on the providers' choice of effort is shown to be positive. A change in the valuation for a service may be caused by either altered preferences of the consumer (substitution effects) or by income effects. It is shown that the optimal levels of effort and the resulting penalties of all providers increase in the valuation of the consumer. This result crucially depends on the assumption that the valuation of the consumer is common knowledge.

This, along with other assumptions of the model, is rather restrictive as compared to practice. Some of the assumptions can be relaxed however without changing the main results related to the optimal effort and penalties.

It is a strong assumption that all bargaining power is concentrated at the consumption level. In reality, even if there is strong competition among service providers, consumers have to comply with the rivalry of other consumers which reduces their bargaining power with respect to the providers. In the model, the welfare-maximizing solution coincides with the private perspective due to risk-neutrality of all participants. As long as all actors are risk-neutral, any redistribution of bargaining power will only affect the prices, as a means of redistributing surplus among the market participants, but not the resulting effort levels and penalties. Risk neutrality of all agents allows for penalties to play the role only of an incentive devise and not of a devise for risk redistribution. It is risk-neutrality because of which there is no difference in the surplus which can be attained from a social and from a private perspective.

Similarly, prices are only affected as a level variable and penalties remain unaffected if the assumption of the one-to-one relationship between services s_i and service providers A_i is relaxed as long as services are assumed to be perfect complements.

An interesting variation of the assumptions which will affect the resulting penalties is the case of limited liability of the service providers. Instead of earning zero profit, $E(\pi_i) \geq 0$, in expectation, this limited liability imposes the requirement of non-negative profits in any contingency.

Far more critical, however, are the assumptions of common knowledge not only with respect to the valuation of the customer but also to the costs of the service providers. These assumptions are necessary in order to separate the different types of information asymmetries which might arise in SLAs along the value chain. In the context of incentivation of service providers ex post of bargaining the service level agreement, common knoweledge of all variables but the effort level, which cannot be directly observed and verified, is assumed. Rather than dealing with hidden *charactersitics* of the bargaining parties, the focus in this chapter is on service providers' *behavior*. In practice, service providers will not have complete information about the consumer's valuation v and other providers' (marginal) costs c_i and will instead apply estimations of consumers' valuation and providers' costs to substitute these values. An even more serious issue, from a practical point of view, might be the identification of the characteristics of the probability functions, $p_i(e_i), \forall i$. These informational restrictions arise not only with respect to the probability distributions of the other service providers' service failure but also with respect to their own probability distribution or their own marginal costs. Unless service providers have access to exhaustive historical data, it cannot be taken for granted that they know the probability distribution of service failure. The probability of service delivery might be approximated based on past observations on effort and resulting service delivery. Marginal costs which are also essential for the calculation of prices and penalties might also be available only as rough approximations.While the results are not meant to be used for the calculation of precise values for the prices and penalties, they give a profound understanding of the underlying effects and show in which direction providers should adjust existing penalties if external conditions change.

A further issue in the chain scenario with consecutive conclusion of SLAs in this chapter is that a service provider interacting only with the immediate neighboring upstream and downstream parties, might not be familiar with the exact number of single services involved in the composition of the complex service. In contrast, the assumption that there is common knowledge of the number of single services and observability of the source of failure is more justified in a scenario in which a central intermediating platform integrates the single services to a complex one and thereby interacts with all involved parties (Filipova-Neumann et al. 2010).

The results presented in this chapter give answers to questions related to moral hazard and external effects between providers acting in service composition platforms. While the proposed model makes simplifying assumptions that limit its direct applicability in practical settings, its focus lies on the behavioral consequences of numerous single services constituting a complex service. An essential characteristic of complex services is that they consist of complementary services. It is shown that complementarities of services and the dependence of service delivery on providers' effort generate external effects and the need for "internalizing" these effect by means of setting the appropriate incentives. Pricing and penalty decisions are shown to implement these incentives.

While some of the assumptions were argued not to be essential for the main results, there are some extensions to the model presented in this chapter which need to be considered separately as they will have substantial impact on the results.

Relaxing the perfect complementarity assumption by means of services that are not perfect complements will render the number of single services which are necessary for the composition of the complex service into an endogenous variable. Thus, complementarity of services can turn into a strategic variable of service providers by means of which they can influence the degree of competition and cooperation. The degree of complementarity which they will choose and the factors which influence their choice are questions to be explored in future research.

With respect to the practical applicability of the model, it should be adjusted to more closely reflect reality. The question which arises when revoking the assumption of common knowledge is how information asymmetries with respect to effort, which are the subject of this chapter, interact with information asymmetries with respect to services' attributes, providers' costs, and consumers' valuations. Under such conditions, the market structure will matter and hence, different scenarios should be considered. Moreover, a model claiming a higher practical applicability cannot ignore the dynamic aspect of bargaining and bidding which raises raises important issues of market design.

References

Archer A, Tardos É (2007) Frugal path mechanisms. ACM Trans Algorithm (TALG) 3(1):1–22

Becker M, Borrisov N, Deora V, Rana OF, Neumann D (2008) Using k-pricing for penalty calculation in grid market. Proceedings of the 41st Annual Hawaii international conference on system sciences (HICSS 2008) pp 97–97

Blau B (2009) Coordination in service value networks. Dissertation

Blau B, Kramer J, Conte T, Dinther C (2009) Service value networks. In: Proceedings of the 2009 IEEE conference on commerce and enterprise computing-volume 00. IEEE Computer Society, Washington, DC, pp 194–201

Crockford, Douglas. JSON: The fat-free alternative to XML. XML 2006, Boston. December 6, 2006.

Fielding R, Taylor R (2002) Principled design of the modern web architecture. ACM Trans Internet Technol (TOIT) 2(2):115–150

Filipova-Neumann L, Michalk W, Blau B, Weinhardt C (2010) Incentivation of service providers for efficient service composition agreements. In: Proceedings of the international conference on e-Commerce, e-Business and e-Service 2010 (ICCBS 2010), Paris, France, pp 1569–1583

Hagel JI (1996) Spider versus spieder. McKinsey Q 1:71–80

Hershberger J, Suri S (2001) Vickrey prices and shortest paths: What is an edge worth? In: Annual Symposium on Foundations of Computer Science, Citeseer, vol 42, pp 252–259

Holmström B (1979) Moral hazard and observability. Bell J Econ 10:74–91

Immorlica N, Karger D, Nikolova E, Sami R (2005) First-price path auctions. In: Proceedings of the 6th ACM conference on electronic commerce. ACM, New York, pp 203–212

Leymann F (2003) Web services: Distributed applications without limits. In Gerhard Weikum, Harald Schöning, Erhard Rahm, editors, BTW 2003, Datenbanksysteme für Business, Technologie und Web, Tagungsband der 10. BTW-Konferenz, 26.-28. Februar 2003, Leipzig. Volume 26 of LNI, pages 2–23, GI, 2003

Macías M, Guitart J (2008) Influence of Reputation in Revenue of Grid Service Providers. 2nd International Workshop on High Performance Grid Middleware (HiPerGRID 2008), Bucharest, Romania, pp. 9–16, ISSN: 2065-0701, November 2008

Nisan N, Ronen A (2001) Algorithmic mechanism design. Games Econ Behav 35(1–2):166–196. doi: 10.1006/game.1999.0790

Papazoglou M (2008) Web services: Principles and technology. Pearson Prentice Hall, Upper Saddle River

Ross SA (1973) The economic theory of agency: The principcal's problem. Am Econ Rev 63:134–139

Shavell S (1979) Risk sharing and incentives in the principal and agent relationship. Bell J Econ 10:55–73

Spulber DF (1999) Market microstructure: Intermediaries and the theory of the firm. Cambridge University Press, Cambridge

Stiglitz JE (1975) Incentives, risk and information: Notes toward a theory of hierarchy. Bell J Econ 6:552–579

Chapter 7
Coordination and Pricing in Service Value Networks: A Mechanism Design Approach

I believe that in the future we may see much more auctioning of services [...]. Services are particularly attractive for auctions because they are in relatively fixed supply – unlike durable goods, one cannot store surpluses or draw down inventory in order to meet fluctuating demand.

Lucking-Reiley (2000)

Weinhardt et al. (2003) and Neumann (2004) state that there is no general mechanism available to fit any possible market setting. In accordance with this statement, it is necessary to present a suitable mechanism designed to fit the underlying field of application. The adequacy of a mechanism depends, amongst others, on the properties of the trading objects. In SVNs, the latter are modular Web service components as well as the composed complex services resulting thereof – whose characteristics were discussed in detail in Chap. 3.

Mechanism design is a subfield of game theory that takes over an engineering perspective (Nisan 2007). The goal of the mechanism is manifested in a so-called social choice function that reflects the design objective. Mechanism design has proven to be a powerful instrument to solve problems that involve self-interested individuals that hold private information (Mas-Colell et al. 1995; Parkes 2001). The main focal point of mechanism design is not what will happen in a specific interaction of various agents, but rather how to tackle the challenge of having a desired outcome in mind and to comprehend which strategic interaction and which setting could lead to a course of action that implements this very outcome (Shoham and Leyton-Brown 2008).

In particular, auction-based approaches are said to be ideally suited to extract distributed valuations in environments that involve the trading of heterogenous services (Smith 1989; Pesendorfer and Swinkels 2000; Jackson 2003a).

C. Weinhardt et al., *Business Aspects of Web Services*,
DOI 10.1007/978-3-642-22447-8_7, © Springer-Verlag Berlin Heidelberg 2011

7.1 Mechanism Design Foundations

The discipline of mechanism design focuses on implementing a preferred system-wide solution to a decentralized optimization problem where self-interested agents act according to their private preferences for different outcomes (Parkes 2001). The agents' private information cannot be verified by some central institution such as a market or platform operator that seeks to achieve certain objectives (Jackson 2003b). Therefore, the goal pursued cannot be solved directly. One has to design a mechanism which establishes a set of incentives, for instance, via side payments to effectively coordinate participants and to eventually enforce the system-wide solution (Nisan and Ronen 2001). Such side payments shall compensate the agents for potential individual disadvantages that arise if the desired result occurs. In other words, the challenge of mechanism design is to implement institutional rules which determine decisions as a function of the information known by the individuals in the economy, thereby ensuring desired events to occur even if participants act strategically in order to maximize their individual utility (Myerson 1988).

A mechanism operator finds itself in a situation where participants act opportunistically and hold private information on their types which cannot be extracted directly. Therefore, the mechanism operator does not only have to define its very goal, but also needs to specify the framework, i.e., the rules, of the game at hand. In a first step, the system-wide goal is defined via a social choice function. It is to select a desired outcome, given the types of all considered agents, that meets the (howsoever selected) favored properties reflecting the ideal of the mechanism operator (Parkes 2001). Agents can be persons, organizations, or technical systems. Regardless of whether an agent is human or not, rationality is one of the basic assumptions, that is, agents want to maximize their own profits. Each agent is endowed with possibly different preferences for different outcomes of a game. The game itself can be interpreted as a set of rules according to which agents can act (Shapley 1953). The type of an agent denotes its private information relating to its preferences with respect to possible outcomes of the game (Jackson 2003b).

In a second step, the above-mentioned rules of the game at hand need to be defined. Such rules are called the *mechanism* and determine the outcome of the game given the strategies of the agents. In more detail, the mechanism is an *outcome rule* which is composed of an *allocation or choice function* and a *transfer or payment function*. The allocation rule defines the distribution over choices given the agents' strategies while the transfer rule determines the (monetary) payments made to or by the agents.[1] Based on its type and the rules of the game, an agent will make decisions on how to act in it. This is referred to as an agents strategy.

Classic mechanism design originates from the seminal contributions provided by Vickrey (1961), Clarke (1971), and Groves (1973), that are namesake to the

[1]Note that the mechanism can only be decomposed into allocation and transfer rule if quasi-linear preferences are assumed (Shoham and Leyton-Brown 2008).

prominent class of Vickrey-Clarke-Groves (VCG) mechanisms. VCG mechanisms are second-price-sealed-bid mechanisms whose focus is on enforcing *incentive compatibility (IC)*. In incentive compatible mechanisms, all agents are maximize their utility by revealing their true type as an equilibrium strategy.[2]

The focus of classic mechanism design is traditionally put upon three design desiderata closely related to IC. IC is actually a vehicle for another crucial desiderata – *allocative efficiency (AE)* – which says that the mechanism' social choice function always determines an optimal outcome which yields a higher valuation for all agents than any other possible alternative outcome, that is, the total value over all agents (i.e., the system's welfare) is maximized. Further, *individual rationality* (IR)[3] assures that agents do not incur losses by participating in the game. *Budget balance* (BB) makes sure that the mechanism distributes and collects the same amount of money to and from the agents,[4] that is, no outside payments are required to realize the mechanism (Shoham and Leyton-Brown 2008).

As academia has proven, one has to tackle impossibilities among above-described desiderata. The most prominent and far-reaching impossibility was shown by Myerson and Satterthwaite (1983). Given quasi-linear preferences, AE, weak BB, and ex interim IR cannot be fulfilled by any mechanism at the same time, regardless if Bayesian Nash IC us fulfilled or not. This insight results in the fact that the named desiderata need to be balanced in some way. For instance, VCG mechanisms can lead to a considerable need to plough money into the mechanism. Likewise, individually rational and budget balanced mechanisms can result in highly inefficient outcomes (McAfee 1992; Barberà and Jackson 1995).

7.2 Service Selection via Auctions

As outlined in Sect. 3.4, forerunners of SVNs are already in their starting blocks. Yet, neither current pricing models nor the fashion of service coordination reflect the agile and distributed nature of the environment the services are traded in. Generally, pricing is static for the most part, at best allowing for price discrimination, for instance, with respect to the volume of services traded (cp. Sect. 5.1.1). Such static pricing does not include situational preferences of service providers.

In related domains, companies have already discovered the potential of using auctions to sell their offerings. Auctions are said to perform particularly well

[2]Note that there can be different underlying equilibrium concepts, for example an equilibrium in dominant strategies or a Bayesian Nash equilibrium. The interested reader is referred to Mas-Colell et al. (1995).

[3]IR can be either based on the agents' expected utility *(ex interim IR)* or their ex post utility *(ex post IR)*.

[4]If net transfers can be made from agents to the mechanism (but not the other way round), one speaks of *weak BB*.

in settings in which heterogenous and intangible trading objects are exchanged (Smith 1989). For example, Google exploits this potential with its advertising service AdWords.[5] In order to price and allocate advertising space, Google applies a generalized second-price auction which includes a quality score to classify advertisements (Varian 2009). Edelman et al. (2007) reports that this auction generated 98% of Google's 2005 revenues.

Recently, the auctioning of services has taken a major step towards mainstream acceptance with Amazon's EC2 Spot Instances.[6] EC2 Web service capacity that is not directly sold via Amazon's traditional price model is offered in an auction: as long as a customer's bid exceeds the current spot price, the auctioned EC2 instance can be run.[7] That way, termination of used instances is rather unpredictable, yet the model supports customers that are flexible as to execution time. The EC2 spot instances are a remarkable example of how service providers exploit (one-sided) auction settings in order to capture quickly changing demand, in Amazon's case, to optimize capacity utilization.

Section 7.2.1 focuses on an analysis of the applied auction setting in SVNs, namely multiattribute reverse auctions, thereby cautiously forgoing a general overview on auction theory. For a comprehensive overview on auction theory, the interested reader is referred to Klemperer (2004) and Milgrom (2004). Customized to SVNs, Sect. 7.2.2 presents the bidding language designed for the mechanism design approaches presented in this chapter.

7.2.1 Procurement Auctions in Service Markets

As introduced in Chap. 3, customers approach an SVN with the objective of buying a complex service that is tailored to their requirements. While it maybe possible to procure simple kinds of services "off-the-shelf" by only negotiating along one dimension – the price – without specifying any attributes that configure and fix quality of service, this is certainly not true for the kind of complex Web services traded in service value networks. Therefore, the auctioning of services in SVNs should be described by means of a *multiattribute procurement auction*.

In *single-sided reverse procurement auctions*, a single buyer – the service customer – receives bids from multiple competing sellers (Bichler 2001; Jap 2002; Klemperer 2004). *Multiattributivity*, as introduced to auction design by Che et al. (1993) and Branco (1997), allows for the negotiation over non-price attributes by referring to multiple features of a single unit (Wellman 2005).

In Web service scenarios, multiattribute approaches allowing for the negotiation over various non-price attributes that constitute a service's configuration is of crucial

---

[5]http://adwords.google.com/.

[6]http://aws.amazon.com/ec2/spot-instances/.

[7]The spot price changes periodically based on actual supply and demand.

importance. Particularly in SVNs, in which complex services are assembled from complementary service offerings, a multitude of different service configurations must be handled and aggregated, adding up to a potentially vast variety of complex services that offer different QoS. For each of these configurations, buyers have a (most likely different) valuation. Procurement auctions determine which service providers are to win the auction, which service configuration has to be provided by them, and which payment has to be made by the service customer. Importantly, multiattribute procurement auctions allow different service providers to compete over both attribute values and price. This is a particularly important feature in the service world in which quality has become the main differentiator thanks to quickly decreasing ICT costs and harsh price competition (Matthyssens and Vandenbempt 1998; Dan et al. 2003; Papazoglou 2008).

Due to their complexity, procurement negotiations on multiple services and on services with multiple attributes have traditionally been conducted manually as request for quotes or via phone negotiations (Bichler et al. 2009). However, in the last decade, electronic auctions in procurement settings have emerged and proven to suit the problem set remarkably well: today powerful computer networks have brought up electronic marketplaces that can handle the procurement of multiattribute services through automated negotiation and the determination of the winning complex service (Bichler and Kalagnanam 2006; Chandrashekar et al. 2007; Bichler et al. 2009).

Beall (2003) surveyed that by as early as 2002, electronic reverse auctions were used for an average of approximately 4% of the polled companies' total spend. Larger companies already use procurement auctions for more than 25% of their spend. These numbers were expected to exhibit distinct growth in the following years. Currently, procurement auctions are mainly applied by companies that seek to optimize their procurement activities, relying on commercial software as for example provided by Ariba[8] or i2 Technologies.[9] Thus, the mechanism operator, or auctioneer, respectively, and the buyer coincide. However, with emerging marketplaces such as SVNs and due to the power of combinatorics of modular Web services and the effects of the long valley, a procurement mechanism offered to "third party buyers" becomes more and more interesting as a rule set and framework to enable transactions. Such third party buyers are customers interested in procuring complex services via a service marketplace.

Academia and practice have brought up different variants of procurement auctions. Besides the multiattribute character, procurement auctions can either be designed for single-unit or for multi-unit settings. In the latter, sellers bid for bundles and/or are given the possibility to negotiate over volumes (Bichler and Kalagnanam 2006; Chandrashekar et al. 2007). This is not the case in SVNs where quality, not quantity is in focus: service customers request only one complex service that is *technically interpreted as a single item*. That is, the winning complex service is in

[8]http://www.ariba.com/about/.

[9]http://www.i2.com/.

fact assembled from the offerings of diverse service providers, yet the set of required service components and its sequence is fixed. Exactly one component service out of each candidate pool will be chosen to add up to the complex service requested. Different customer valuations for bundles of services, for instance, one bundle comprising a service from candidate pools Y^1 and Y^2 and a second bundle including services from Y^2 and Y^3, are not applicable since the buyer will only procure a complex service that includes a defined sequence of services (e.g., consisting of the candidate pools Y^1, Y^2, and Y^3, in exactly this sequence), otherwise its valuation is zero. Thus, a variety of different services is available in the same market, however, the customer's valuation and its bidding language is less complicated than in a typical combinatorial auction setting.[10] Transferred to SVNs, multiattribute auctions allow for expressing preferences over configurations of a specific complex service rather than over bundles of services as supported by combinatorial approaches (Engel et al. 2006).

In general, the reverse character of procurement auctions with multiple units being allocated can lead to situations where the valuation on the buyer-side (i.e., the service customer in the SVN setting) is less than the total payments to be made to the sellers (i.e., service providers) (Kothari et al. 2005). Parkes and Kalagnanam (2005) state that the multiattribute allocation problem with a single buyer belongs to settings with two-sided private information. This complicates the determination of the allocated service providers since, besides the private seller types, market clearance additionally depends on the revealed preferences of the buyer.

7.2.2 *Bidding Language: The Scoring Function*

As a formalization of information objects which are exchanged during the auction conduction a bidding language is introduced that is based on bidding languages for products with multiple attributes as discussed in Engel et al. (2006). The formalization is aligned to multi-attribute auction theory as presented in Parkes and Kalagnanam (2002a) and Ronen and Lehmann (2005) and assures compliance with the WS-Agreement specification (Andrieux et al. 2004) in order to enable realization in decentralized environments such as the Web.

A complex service – represented by a path $F_k \in F$ – is characterized by a configuration $\mathscr{A}_{F_k}$. The importance of certain attributes and prices of a requested complex service is idiosyncratic and depends on the preferences of the requester. The requesters' preferences are represented by a scoring function $\mathscr{S}$ of the form:

$$\mathscr{S}(\mathscr{A}_{F_k}) = \left(\sum_{l=1}^{m} \lambda_l \| \mathscr{A}_{F_k}^l \| \right) \tag{7.1}$$

[10]Combinatorial auctions are not assessed in more detail in this book. For a comprehensive overview of this kind of auction setting, please refer to Cramton et al. (2006).

The scoring function $\mathscr{S}$ represents the requesters' preferences for a configuration $\mathscr{A}_{F_k}$ of the complex service represented by F_k analog to the definition of scoring rules in Asker and Cantillon (2008). It maps the configuration of a complex service to a value representing the requester's score such that $\mathscr{S} : \mathscr{A} \rightarrow [0;1]$. The scoring function is determined by a vector of weights $\Lambda = (\lambda_1, \ldots, \lambda_m)$ with $\sum_{l=1}^{m} \lambda_l = 1$ that defines the requester's preferences of each attribute type $l \in \mathscr{L}$. The configuration $\mathscr{A}_{F_k}$ of the complex service F_k is constituted by the aggregation of all attribute values of contributing services with incoming edges on the path F_k such that

$$\mathscr{A}_{F_k} = (\mathscr{A}_{F_k}^1, \ldots, \mathscr{A}_{F_k}^m) \text{ with } \mathscr{A}_{F_k}^l = \bigoplus_{e_{ij} \in F_k} a_j^l \tag{7.2}$$

The aggregation operation $\bigoplus$ for attribute values depends on their type (e.g., the attribute type encryption is aggregated using a Boolean AND operator whereas response time is aggregated by a sum operator). Table 7.1 shows different types of aggregation functions for sample multiple attribute types.

The list of aggregation operations in Table 7.1 only shows a rather trivial subset of possible and practical aggregation operations for different quality aspects of services and is not exhaustive. The bidding language also supports rich semantic approaches towards more complex aggregation operations in order to deal with various non-functional service attributes. For example, services are capable of different types of encryption algorithms and a requester prefers asymmetric ciphers, semantic subsumption can be used to evaluate if a complex service fulfils the requester's requirements and therefore to determine the score.

7.3 The Complex Service Auction

Auctions have proven to perform quite well in situations where intangible and heterogenous entities are traded Smith (1989). Furthermore, valuations are hard to determine for single and especially value-added complex services as the value of the service's outcome highly depends on the customer's preferences for which current pricing models do not account. Auctions are predestinated to *aggregate information*

Table 7.1 Aggregation operations for different attribute types

Attribute type	Aggregation
$l \in \mathscr{L}$	$\bigoplus_{e_{ij} \in F_k \mid j \neq v_f} a_j^l$
Response time (rt)	$\sum_{e_{ij} \in F_k \mid j \neq v_f} a_j^{rt}$
Encryption type (et)	$\bigwedge_{e_{ij} \in F_k \mid j \neq v_f} a_j^{et}$
Error rate (er)	$\max_{e_{ij} \in F_k \mid j \neq v_f} a_j^{er}$
Throughput (tp)	$\min_{e_{ij} \in F_k \mid j \neq v_f} a_j^{tp}$
Probability of default (pd)	$1 - \prod_{e_{ij} \in F_k \mid j \neq v_f} \left(1 - a_j^{pd}\right)$

from distributed parties which results in an aggregated valuation (Pesendorfer and Swinkels 2000; Jackson 2003a). Without prior knowledge about the valuations of each participant, auctions can provide suitable incentives to make truth-revelation an equilibrium strategy and therefore automatically *aggregate necessary information* from self-interested participants to determine adequate prices for complex services.

This section introduces the Complex Service Auction based on Blau (2009) which enables the efficient trade of complex services that are build through the composition of distributed services in service value networks. The design of the auction mechanism exposes valuable technical and economic properties that are illustrated and discussed in the remainder of this section.

7.3.1 Service Request and Service Offer

A complex service – represented by a path f – is characterized by a configuration $\mathscr{A}_f$. The importance of certain attributes and prices of a requested complex service is idiosyncratic and depends on the preferences of the requester. The requesters' preferences are represented by a scoring function $\mathscr{S}(\mathscr{A}_f)$. The scoring function represents the requesters' preferences for a configuration $\mathscr{A}_f$ of the complex service represented by f analog to the definition of scoring rules in Asker and Cantillon (2008). It maps the configuration of a complex service to a value representing the requester's score

Having defined how the score for certain outcomes is computed based on the requester's preferences, a specification of the willingness to pay is introduced that determines the rate of substitution between score and price. Let $\mathscr{T}_f = \sum_{s \in S} t^s$ represent the sum of all monetary transfers to service providers, i.e., the overall price of the complex service denoted by f. Hence, the requester's utility gained from purchasing a complex service specified by a path f with a configuration $\mathscr{A}_f$ evolves as follows:

$$\mathscr{U}_f^R(\alpha, \Lambda, \mathscr{A}_f, \mathscr{T}_f) = \alpha \mathscr{S}(\mathscr{A}_f) - \mathscr{T}_f \tag{7.3}$$

The factor α represents the requester's willingness to pay for a "perfect" configuration $\mathscr{A}_f$ with score $\mathscr{S}(\mathscr{A}_f) = 1$ based on reported preferences. In other words α defines the individual substitution rate between quality and price such that the requester is indifferent between an increase of 1 score unit and α monetary units. Incorporating that information, a service request for a multidimensional complex service is defined as follows:

Definition 7.1 (Multidimensional Service Request). A multidimensional service request for a complex service is a vector of the form:

$$R := (\mathscr{Y}, \alpha, \Lambda) \tag{7.4}$$

such that $\mathscr{Y} = (Y_1, \ldots, Y_K)$ represents all candidate pools with the service value network, i.e., necessary information for each service provider about preceding service offers.[11] The maximum willingness to pay for a configuration that yields a score of 1 is denoted by α. The set of weights Λ represents the requesters' preferences for different attribute types $l \in \mathscr{L}$.

After the bidding language for requesters is specified, a notation for the provider side can be defined. A multidimensional service offer consists of an announced service configuration A_j and a corresponding price p_{ij} that a service provider wants to charge for the service j being invoked depending on the predecessor service i. An offer bid $b_{ij} = (A_j, p_{ij})$ is a service offer for invocation of service j as a successor of service i. A service provider s announces a matrix of bids $B^s \in \mathscr{B}$ for all incoming edges to every service it owns:

Definition 7.2 (Multidimensional Service Offer). A multidimensional service offer is a matrix of bids of the form:

$$B^s := \begin{cases} b_{ij} = (A_j, p_{ij}), & i \in \tau(j), j \in \sigma(s) \\ (\bar{A}_j, -\infty), & \text{otherwise} \end{cases} \tag{7.5}$$

with $\tau(v)$ denotes the set of all predecessor services to service v with $\tau : V \to V$ and $\sigma(s)$ the set of all services owned by service provider s. $\bar{A}_j$ is an arbitrary service configuration.

7.3.2 *Winner Determination*

Let $\mathscr{U}_f$ denote the overall utility of path f based on the reported types. Let further $\mathscr{P}_f$ be the sum of all price bids for allocated service offers on the path f such that $\mathscr{P}_f = \sum_{e_{ij} \in f} p_{ij}$. The allocation function $o : \mathscr{B} \to F$ maps the service providers' bids $B \in \mathscr{B}$ – their reported types – to a feasible path from source to sink $f^* \in F$[12] such that:

$$o(B) := \underset{f \in F}{\operatorname{argmax}}\, \mathscr{U}_f = \underset{f \in F}{\operatorname{argmax}}\, \left(\alpha \mathscr{S}(\mathscr{A}_f) - \mathscr{P}_f\right) \tag{7.6}$$

Having defined an allocation function to perform a desired social choice that selects a set of edges within G that determine the instance of the complex service, a function that specifies monetary transfers to service providers has to be designed.

[11] Note that there are no preceding service offers for services v with $v \in Y_1$.

[12] For the sake of simplicity, the expression "allocated service offer" means that this service offer has an incoming edge that is entailed in the allocated set of edges f^*. Analogously, the expression "allocated service provider" means that a service provider owns at least one "allocated service offer."

Let $\mathcal{U}^*$[13] denote the overall utility of the allocated path meaning the utility of the path f^*, which maximizes the overall utility. Furthermore, let $\mathcal{U}^*_{-s}$ denote the overall utility of a path f^*_{-s} that yields the maximum welfare in a reduced graph G_{-s} without every service owned by service provider s and without incoming and outgoing edges of these service offers, i.e., the complex service instance that maximizes welfare in an service value network without service provider s's participation.

Definition 7.3 (Critical Value). The *critical value* $\Delta t^{crit,s}$ of a service provider s represents its contribution to the system as the difference between the overall utility $\mathcal{U}^*$ in the complete graph and the overall utility in the reduced graph $\mathcal{U}^*_{-s}$ without service offers owned by service provider s and incoming and outgoing edges of these services such that

$$\Delta t^{crit,s} = \mathcal{U}^* - \mathcal{U}^*_{-s} \tag{7.7}$$

7.3.3 *Auction Process and Realization*

The auction conduction is divided in two main phases: a *solicitation phase* and the actual *auction phase* as depicted in Fig. 7.1.

The solicitation phase serves as an initial screening phase regarding the service request and potential service provider candidates to be invited to participate in the auction. The service requester sends a *complex service solicitation* to the service intermediary which initiates the coordination process. The complex service solicitation specifies required modularized functionality which determines the candidate pools that are sequentially involved in the production of the complex service requested.

Based on this information, the service intermediary reasons about potential service providers to be invited to participate in the auction phase. There are different forms of finding and defining suitable participants. The service intermediary can step into the role of pushing the invitation process using e.g., a registry to find adequate service providers. It is also possible to reverse the roles in such a lookup scenario, meaning that potential participants are proactively searching for suitable coordination services provided by a service intermediary. Potential participants could also subscribe to a notification service – analogue to the observer design pattern – in order to automatically be informed if an adequate auction service is available.

Having defined the set of potential service providers to participate in the auction phase, the service intermediary sends out the complex service solicitation and additional information as an invitation to the candidates. This information enables

[13] For the reader's convenience, the notion $\mathcal{U}^*$ is short for $\mathcal{U}_{o(B)}$ which denotes the overall utility of the path f^* allocated by $o(B)$ based on service providers' bids.

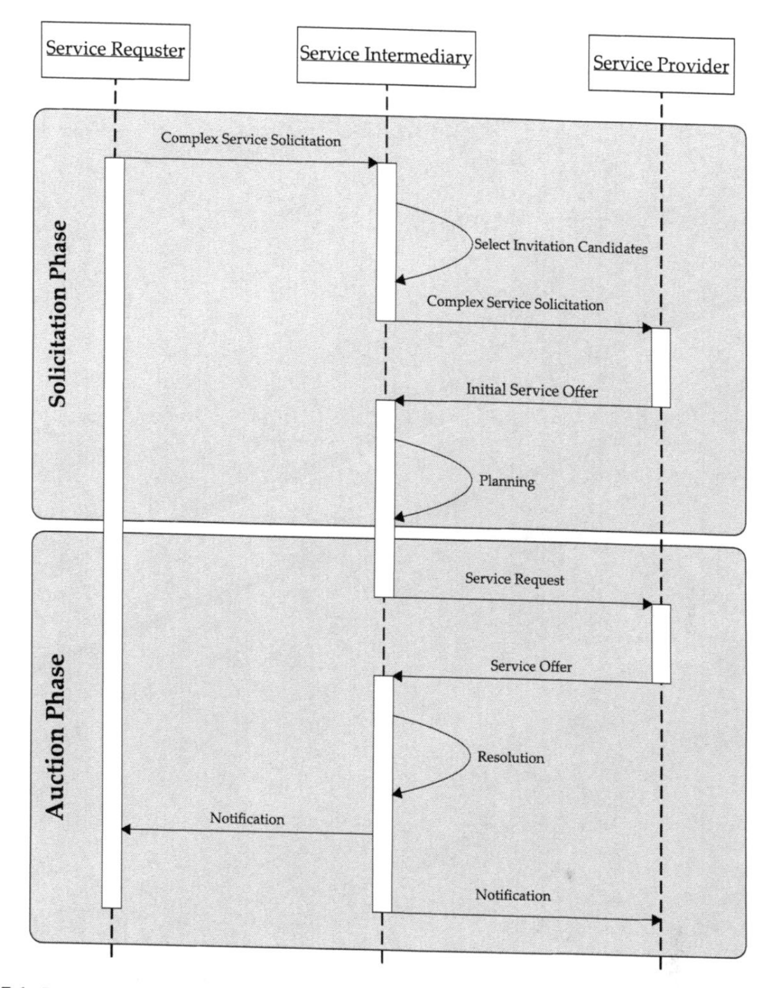

Fig. 7.1 Process model of the CSA

service providers to register their service offerings to be part of the service value network and to be considered in the auction phase by sending *initial service offers*.

Combining the information about the complex service solicitation and the initial service offers from service providers, the service intermediary *plans* the topology of the service value network and proceeds its virtual formation. This step concludes the solicitation phase and lays the basis to the actual auction phase.

The auction phase embodies the central coordination process to allocate and price complex services. Messages and information objects exchanged during the auction conduction are fully specified according to the bidding language. The topology information about the service value network as well as the requester's preferences and willingness to pay is sent as a *service request* to registered service providers. Having received the requester's information, the service providers

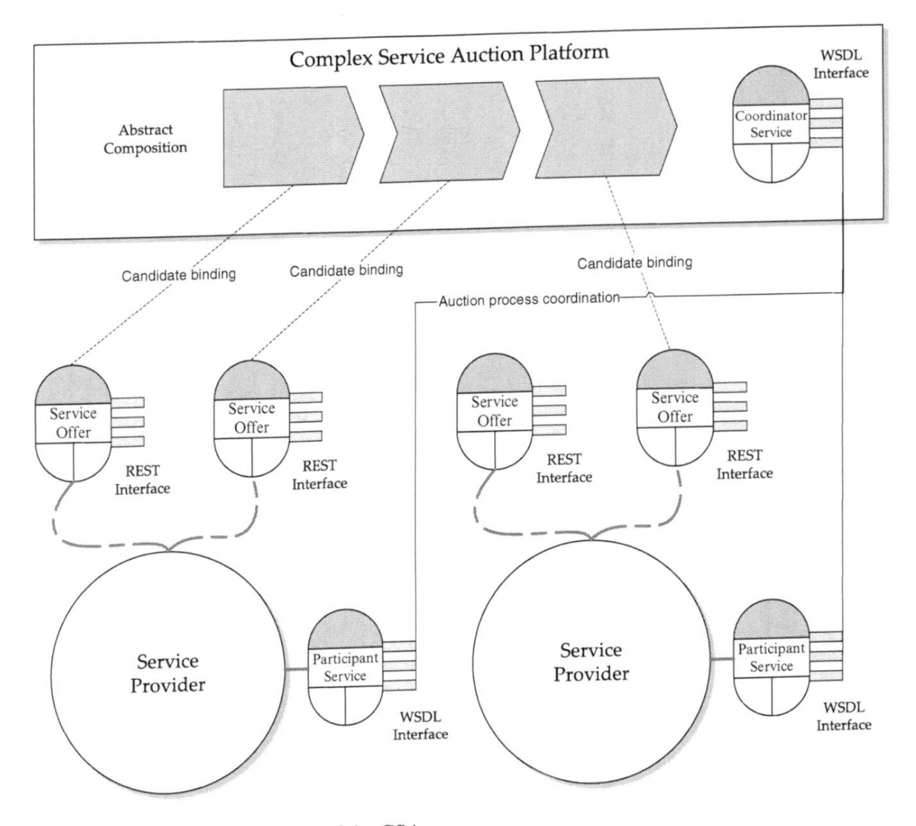

Fig. 7.2 Architectural overview of the CSA

privately submit their *service offers* to the service intermediary. Having collected necessary information from requester and provider side, the service intermediary *resolves* the auction by computing the winner determination and resulting monetary transfers. The auction process concludes with notifications about the final outcome and corresponding transfers sent to the service requester and the service providers.

Providing an architectural overview, Fig. 7.2 shows service providers that intent to participate in the auction, their service offers which are realized in a lightweight manner and necessary big Web services that enable the overall coordination of the auction process.

The Complex Service Auction platform as the central coordination unit communicates with potential participants via a *coordinator service* implemented as a Web service with a WSDL interface. Analogously, each service provider exposes a *participant service* for the message exchange with the coordinator. After the coordination phase is completed, concrete candidate service instances are bound to each step in the abstract composition in a lightweight manner leveraging the simplicity of REST/HTTP interfaces. The final composition embodies the outcome of the coordination process in the form of a concrete complex service instance.

7.3.4 Solving the Winner Determination Problem

This section provides an in-depth analysis of the *ComputeAllocation* algorithm which performs the winner determination in the complex service auction. Special challenges that result from aggregation operations such as min and max as well as Boolean operations which are used in the context of semantic QoS extensions are outlined and adequate remedies are discussed. The procedure of the algorithm is illustrated stepwise by means of an extensive example. Furthermore, this section introduces a prototypical implementation of a service value network planner tool and an agent-based simulation tool to analyze the complex service auction.

From an algorithmic mechanism design perspective, computational feasibility according is a central desideratum in order to implement the mechanism in an online system which requires on-the-fly computation at run-time.

It is well-known that solving the winner determination problem in general combinatorial auctions is $\mathcal{NP}$-complete. Focusing on finding efficient computational approaches, several algorithms have been proposed to solve the winner determination problem (Papadimitriou and Steiglitz 1998; Rothkopf et al. 1998; Sandholm et al. 2005).

The solution to the allocation problem can be compute in polynomial time using well-known graph algorithms to determine the "shortest" path within a network such as the Dijkstra algorithm (Dijkstra 1959).

According to the underlying payment scheme the allocation must be computed twice for each allocated service offer – based on the graph with the service offerings of the service provider receiving the payment and without its participation. In the second case, the graph can be preprocessed and reduced by all service offerings owned by the service provider that receives the payment. After the reduction the allocation can be computed accordingly which yields the same time complexity.

Nevertheless, the extension of the complex service auction with respect to complex QoS aggregation using also aggregation operations that require complete information about predecessors' attribute values – memory-dependent attribute types – such as min, max, and Boolean operations may result in sub-optimal solutions using the traditional Dijkstra algorithm. Analogue to the problem of negative edge weights which is well-known in literature (Dijkstra 1959), memory-dependent operations may result in non-monotone utility characteristics. Such behavior conflicts with the main procedure of the Dijkstra algorithm, that is, it truncates a sub-path which is directly dominated by another sub-path that intersects it at the point of intersection. Considering an attribute type *encryption* which is aggregated by a Boolean AND operation. A sub-path f_s^1 dominates another sub-path f_s^2 as it yields a higher utility which results from an aggregated value for encryption of TRUE. In case both sub-paths intersect at a certain node, the Dijkstra algorithm only considers f_s^1 and drops f_s^2 as f_s^1 yields a higher overall utility so far. Nevertheless, this might be error prone if the subsequent service offer does not support encryption which leads to an aggregated encryption value for f_s^1 of FALSE. Hence, the former decision of dropping f_s^2 might have been incorrect since now both sub-paths are not encrypted and f_s^2 might dominate f_s^1 in price.

To overcome illustrated shortcomings of the Dijkstra algorithm, Algorithm 7.1 accounts for attribute types which are aggregated by memory-dependent operations always yielding an optimal solution.

In order to describe the procedure of the *ComputeAllocation* algorithm and its complexity, Algorithm 7.1 is divided into three parts, namely the *initialization phase* (lines 1–6), the *main phase* (lines 7–16) and the *consolidation phase* (lines 17–18).

Initialization phase. In the initialization phase, required variables are initialized and set to their starting values. In contrary to the traditional Dijkstra algorithm, the *ComputeAllocation* algorithm visits every node within the graph which is equal to the worst-case behavior of a Dijkstra search. Therefore, the node queue Q entails all nodes $u \in V$ ordered by the sequence of the candidate pools in the network such that *getNodesPoolWise(V)* $= (u_1^1, \ldots, u_{|Y_1|}^1, \ldots, u_1^K, \ldots, u_{|Y_K|}^K)$[14] with $\{u_1^1, \ldots, u_{|Y_1|}^1\} = Y_1$ and $\{u_1^K, \ldots, u_{|Y_K|}^K\} = Y_K$. The function *getNonMonotoneStates*(u) retrieves all possible combinations of memory-dependent attribute values of service offer u. Exemplary, if service offer u is only characterized by an encryption attribute type with boolean values, hence *getNonMonotoneStates*$(u) = \{\texttt{TRUE}, \texttt{FALSE}\}$. Let the set W entail all possible

Algorithm 7.1 ComputeAllocation

Require: V, E, B
1: $Q \leftarrow$ *getNodesPoolWise*(V)
2: **for all** $u \in Q$ **do**
3:     *states*$[u] \leftarrow$ *getNonMonotoneStates*(u)
4:     **for all** $w \in$ *states*$[u]$ **do**
5:         *utility*$[u][w] \leftarrow -\infty$
6:         *path*$[u][w] \leftarrow \emptyset$
7: **while** *getNextNode*$(Q) \neq$ *null* **do**
8:     $u \leftarrow$ *getNextNode*(Q)
9:     *removeNode*(u, Q)
10:     **for all** $v \in$ *getSuccesors*(u, E) **do**
11:         **for all** $w \in$ *states*$[u]$ **do**
12:             $\bar{w} \leftarrow$ *computeState*(w, e_{uv}, B)
13:             *altUtility* $\leftarrow$ *computeUtility*(*path*$[u][w] \cup \{e_{uv}\}, B)$
14:             **if** *altUtility* $>$ *utility*$[v][\bar{w}]$ **then**
15:                 *utility*$[v][\bar{w}] \leftarrow$ *altUtility*
16:                 *path*$[v][\bar{w}] \leftarrow$ *path*$[u][w] \cup \{e_{uv}\}$
17: $w^* \leftarrow \text{argmax}_{w \in states[v_f]}$ (*utility*$[v_f][w]$)
18: **return** *path*$[v_f][w^*]$

[14]The order within each candidate pool is not important.

states after aggregation, then the time complexity of the initialization phase is $\mathcal{O}(|V| \cdot |W|)$.

Main phase. In the main phase, the algorithm iterates over all nodes in Q and removes each node after processing until there is no entry left in the queue. Each successor v of the current node u is evaluated for all states of u. The utility of the sub-path including v is computed based on the overall utility $\mathcal{U}_f$. These alternatives are compared to the current utility entry for node v and updated in case of improvement. The variables *utility* and *path* store for each node u and each state the highest utility and the corresponding path respectively. Traversing all successors of every node in Q, the *ComputeAllocation* algorithm processes every edge in the main phase and compares every state of each node. This leads to a time complexity of the main phase of $\mathcal{O}(|E| \cdot |W|)$.

Consolidation phase. After the main part has terminated once Q is empty, i.e., all nodes have been processed, the consolidation phase evaluates the results. The path from source to sink is analyzed and the state s^* that maximizes the overall utility is determined. Based on this state the final allocation $path[v_f][s^*]$ is returned. Implemented as a linear search, the consolidation phase yields a time complexity of $\mathcal{O}(|W|)$.

The time complexity of the *ComputeAllocation* algorithm consisting of the initialization phase, the main phase and the consolidation phase evolves as $\mathcal{O}(|V| \cdot |W| + |E| \cdot |W| + |W|)$. Assuming a worst case number of edges with respect to the number of nodes $|E|$ can be substituted by $(\frac{|V|-2}{2})^2 + (|V|-2)$. This leads to an overall complexity of $\mathcal{O}(|W| \cdot |V|^2)$. The time complexity regarding the number of service offers and connecting edges, the number of paths respectively, is polynomial which means that the algorithms run-time is robust with respect to a changing number of participants and feasible complex service instances. In contrary to the $\mathcal{NP}$-complete complexity in general combinatorial auctions, this is a valuable achievement that enables the conduction of the complex service auction in online systems.

Nevertheless, with respect to the number of memory-dependent attribute types and the number of their discrete values, the computational complexity is exponential (e.g., assuming N Boolean attribute types, $|W| = 2^N$). From a domain-specific perspective, the impact of this theoretical result is rather weak, as the number of states that have to be iterated by the algorithm decreases rapidly in the average case. Figure 7.3 illustrates the run-time performance of the *ComputeAllocation* algorithm in a scenario with 100 service offers in 10 candidate pools (cp. Fig. 7.3a) and 1,000 service offers in 100 candidate pools (cp. Fig. 7.3b). The service value network is assumed to be fully connected which means that each service offer has the maximum number of incoming edges which results in the maximum number of feasible paths from source to sink. The algorithm's performance is evaluated dependent on the number of memory-dependent attribute types. Attribute types are assumed to be Boolean and their values are uniformly distributed for each service offer. Although the theoretical worst case analysis of the computational complexity is exponential with respect to the number N of memory-dependent attribute types ($\mathcal{O}(2^N)$), the average case with boolean attribute types results in a linear increasing computation

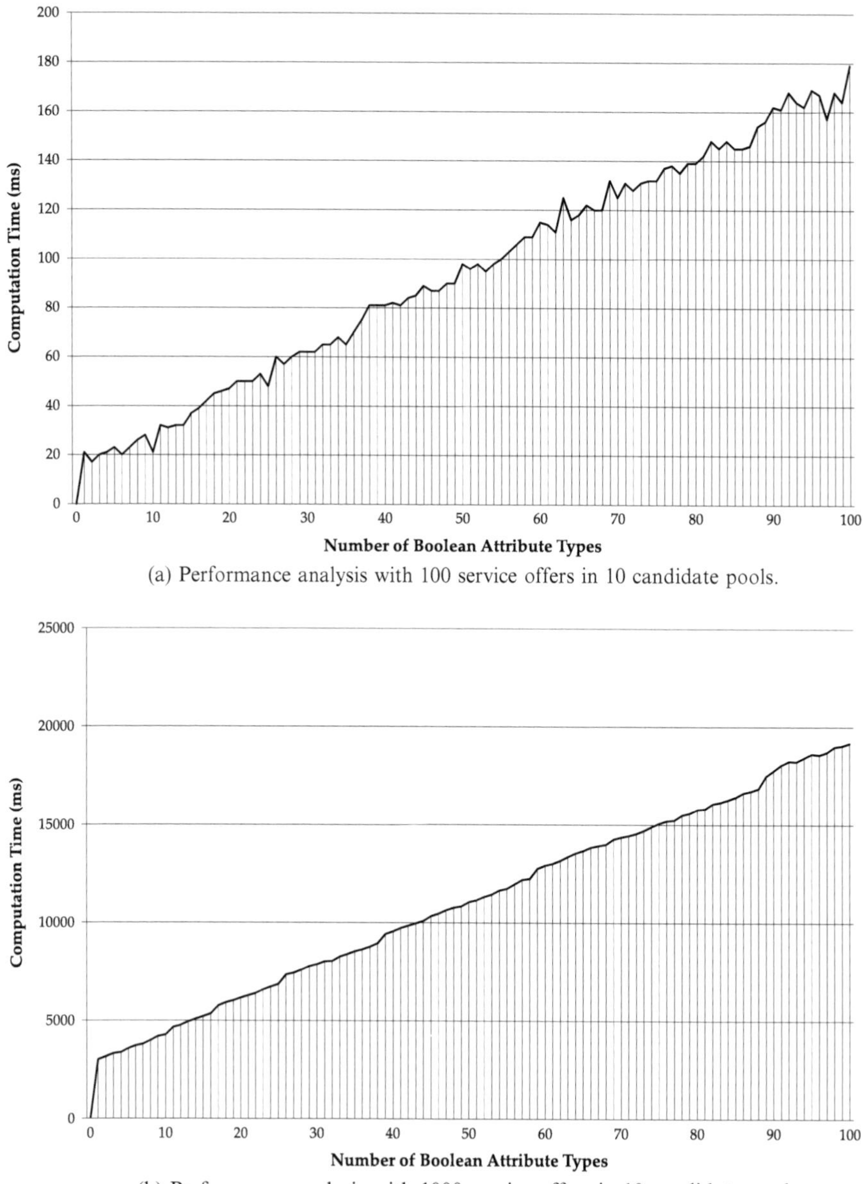

(a) Performance analysis with 100 service offers in 10 candidate pools.

(b) Performance analysis with 1000 service offers in 10 candidate pools.

Fig. 7.3 Performance analysis of the *ComputeAllocation* algorithm

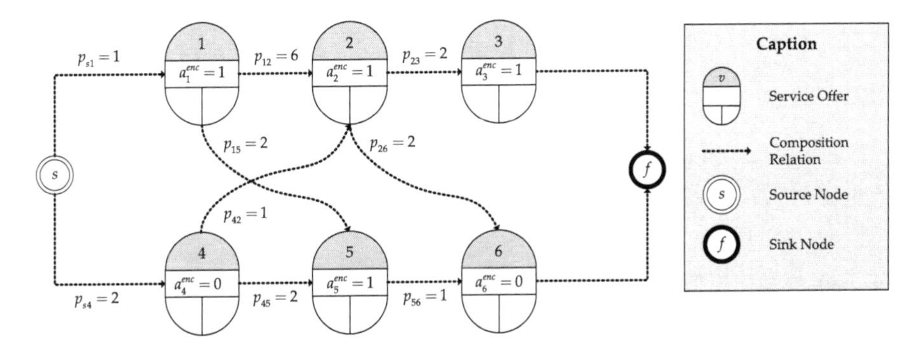

Fig. 7.4 Service value network with service offers exposing memory-dependent attribute types

time. The *ComputeAllocation* algorithm quickly solves the winner determination problem even for huge instances which satisfies computational tractability.

Example 7.1 (Allocation Computation with Memory-dependent QoS). This example illustrates the procedure of the *ComputeAllocation* algorithm in a stepwise manner based on the service value network as depicted in Fig. 7.4.

The service value network consists of six service offers $V = \{1, 2, 3, 4, 5, 6\} \cup \{s, f\}$. Each service offer u is unambiguously configured through a boolean attribute value a_u^{enc} for the attribute type encryption whereas $1 \equiv$ TRUE and $0 \equiv$ FALSE. Values on incoming edges p_{ij} represent price bids of service providers. It is assumed that the service requester's willingness to pay $\alpha\mathscr{S}(\mathscr{A}_f)$ for a complex service depending on its QoS characteristics $\mathscr{A}_f$ evolves as

$$\alpha\mathscr{S}(\mathscr{A}_f) = \begin{cases} 15, & \text{if } \mathscr{A}_f = 1 \\ 12, & \text{if } \mathscr{A}_f = 0 \end{cases}$$

Table 7.2 illustrates the algorithm's procedure to find an optimal allocation based on the allocation function accounting for the memory-dependent attribute type encryption representing the QoS of service offers.

In the last step when node f is processed, the optimal path given a not encrypted complex service results as $f^*_{\text{FALSE}} = \{e_{s1}, e_{15}, e_{56}, e_{6f}\}$ and yields an overall utility of $\mathscr{U}_{f^*_{\text{FALSE}}} = 8$. Given a encrypted complex service, the optimal allocation is $f^*_{\text{TRUE}} = \{e_{s1}, e_{12}, e_{23}, e_{3f}\}$ with an overall utility of $\mathscr{U}_{f^*_{\text{TRUE}}} = 6$. Thus, the state $s^* =$ FALSE yields an optimal path $f^* = \{e_{s1}, e_{15}, e_{56}, e_{6f}\}$ that maximizes the system's overall utility $\mathscr{U}^* = 8$.

Table 7.2 Allocation computation stepwise procedure example

Node			-	s	1	4	2	5	3	6	f
Q			$\{s,1,4,2,5,3,6,f\}$	$\{1,4,2,5,3,6,f\}$	$\{4,2,5,3,6,f\}$	$\{2,5,3,6,f\}$	$\{5,3,6,f\}$	$\{3,6,f\}$	$\{6,f\}$	$\{f\}$	$\emptyset$
utility	s	TRUE	**15**	**15**	**15**	**15**	**15**	**15**	**15**	**15**	**15**
path			$\emptyset$	$\emptyset$	$\emptyset$	$\emptyset$	$\emptyset$	$\emptyset$	$\emptyset$	$\emptyset$	$\emptyset$
utility	s	FALSE	**12**	**12**	**12**	**12**	**12**	**12**	**12**	**12**	**12**
path			$\emptyset$	$\emptyset$	$\emptyset$	$\emptyset$	$\emptyset$	$\emptyset$	$\emptyset$	$\emptyset$	$\emptyset$
utility	1	TRUE	$-\infty$	**14**	**14**	**14**	**14**	**14**	**14**	**14**	**14**
path			$\emptyset$	$\{e_{s1}\}$	$\{e_{s1}\}$	$\{e_{s1}\}$	$\{e_{s1}\}$	$\{e_{s1}\}$	$\{e_{s1}\}$	$\{e_{s1}\}$	$\{e_{s1}\}$
utility	1	FALSE	$-\infty$	$-\infty$	$-\infty$	$-\infty$	$-\infty$	$-\infty$	$-\infty$	$-\infty$	$-\infty$
path			$\emptyset$	$\emptyset$	$\emptyset$	$\emptyset$	$\emptyset$	$\emptyset$	$\emptyset$	$\emptyset$	$\emptyset$
utility	2	TRUE	$-\infty$	$-\infty$	**8**	**8**	**8**	**8**	**8**	**8**	**8**
path			$\emptyset$	$\emptyset$	$\{e_{s1},e_{12}\}$	$\{e_{s1},e_{12}\}$	$\{e_{s1},e_{12}\}$	$\{e_{s1},e_{12}\}$	$\{e_{s1},e_{12}\}$	$\{e_{s1},e_{12}\}$	$\{e_{s1},e_{12}\}$
utility	2	FALSE	$-\infty$	$-\infty$	$-\infty$	**9**	**9**	**9**	**9**	**9**	**9**
path			$\emptyset$	$\emptyset$	$\emptyset$	$\{e_{s4},e_{42}\}$	$\{e_{s4},e_{42}\}$	$\{e_{s4},e_{42}\}$	$\{e_{s4},e_{42}\}$	$\{e_{s4},e_{42}\}$	$\{e_{s4},e_{42}\}$
utility	3	TRUE	$-\infty$	$-\infty$	$-\infty$	$-\infty$	**6**	**6**	**6**	**6**	**6**
path			$\emptyset$	$\emptyset$	$\emptyset$	$\emptyset$	$\{e_{s1},e_{12},e_{23}\}$	$\{e_{s1},e_{12},e_{23}\}$	$\{e_{s1},e_{12},e_{23}\}$	$\{e_{s1},e_{12},e_{23}\}$	$\{e_{s1},e_{12},e_{23}\}$
utility	3	FALSE	$-\infty$	$-\infty$	$-\infty$	$-\infty$	**7**	**7**	**7**	**7**	**7**
path			$\emptyset$	$\emptyset$	$\emptyset$	$\emptyset$	$\{e_{s4},e_{42},e_{23}\}$	$\{e_{s4},e_{42},e_{23}\}$	$\{e_{s4},e_{42},e_{23}\}$	$\{e_{s4},e_{42},e_{23}\}$	$\{e_{s4},e_{42},e_{23}\}$
utility	4	TRUE	$-\infty$	$-\infty$	$-\infty$	$-\infty$	$-\infty$	$-\infty$	$-\infty$	$-\infty$	$-\infty$
path			$\emptyset$	$\emptyset$	$\emptyset$	$\emptyset$	$\emptyset$	$\emptyset$	$\emptyset$	$\emptyset$	$\emptyset$
utility	4	FALSE	$-\infty$	**10**	**10**	**10**	**10**	**10**	**10**	**10**	**10**
path			$\emptyset$	$\{e_{s4}\}$	$\{e_{s4}\}$	$\{e_{s4}\}$	$\{e_{s4}\}$	$\{e_{s4}\}$	$\{e_{s4}\}$	$\{e_{s4}\}$	$\{e_{s4}\}$
utility	5	TRUE	$-\infty$	$-\infty$	**12**	**12**	**12**	**12**	**12**	**12**	**12**
path			$\emptyset$	$\emptyset$	$\{e_{s1},e_{15}\}$	$\{e_{s1},e_{15}\}$	$\{e_{s1},e_{15}\}$	$\{e_{s1},e_{15}\}$	$\{e_{s1},e_{15}\}$	$\{e_{s1},e_{15}\}$	$\{e_{s1},e_{15}\}$
utility	5	FALSE	$-\infty$	$-\infty$	$-\infty$	**8**	**8**	**8**	**8**	**8**	**8**
path			$\emptyset$	$\emptyset$	$\emptyset$	$\{e_{s4},e_{45}\}$	$\{e_{s4},e_{45}\}$	$\{e_{s4},e_{45}\}$	$\{e_{s4},e_{45}\}$	$\{e_{s4},e_{45}\}$	$\{e_{s4},e_{45}\}$
utility	6	TRUE	$-\infty$	$-\infty$	$-\infty$	$-\infty$	$-\infty$	$-\infty$	$-\infty$	$-\infty$	$-\infty$
path			$\emptyset$	$\emptyset$	$\emptyset$	$\emptyset$	$\emptyset$	$\emptyset$	$\{e_{s1},e_{12},e_{26}\}$	$\emptyset$	$\emptyset$
utility	6	FALSE	$-\infty$	$-\infty$	$-\infty$	$-\infty$	**7**	**8**	**8**	**8**	**8**
path			$\emptyset$	$\emptyset$	$\emptyset$	$\emptyset$	$\{e_{s4},e_{42},e_{26}\}$	$\{e_{s1},e_{15},e_{56}\}$	$\{e_{s1},e_{15},e_{56}\}$	$\{e_{s1},e_{15},e_{56}\}$	$\{e_{s1},e_{15},e_{56}\}$
utility	f	TRUE	$-\infty$	$-\infty$	$-\infty$	$-\infty$	$-\infty$	$-\infty$	**6**	**6**	**6**
path			$\emptyset$	$\emptyset$	$\emptyset$	$\emptyset$	$\emptyset$	$\emptyset$	$\{e_{s1},e_{12},e_{23},e_{3f}\}$	$\{e_{s1},e_{12},e_{23},e_{3f}\}$	$\{e_{s1},e_{12},e_{23},e_{3f}\}$
utility	f	FALSE	$-\infty$	$-\infty$	$-\infty$	$-\infty$	$-\infty$	$-\infty$	**7**	**8**	**8**
path			$\emptyset$	$\emptyset$	$\emptyset$	$\emptyset$	$\emptyset$	$\emptyset$	$\{e_{s4},e_{42},e_{23},e_{3f}\}$	$\{e_{s1},e_{15},e_{56},e_{6f}\}$	$\{e_{s1},e_{15},e_{56},e_{6f}\}$

7.3.5 *Summary*

The complex service auction enables the coordination of value generation in service value networks which requires that it is on the one hand theoretically sound and on the other hand applicable in the context of electronic services and their composition. It is a well-known result from algorithmic or computational mechanism design (Nisan and Ronen 2001; Dash et al. 2003) and market engineering (Weinhardt et al. 2003; Neumann 2004) that these theoretical and practical goals are oftentimes conflicting which requires reasonable solutions regarding these trade-offs to satisfy the requirements upon a suitable mechanism in a certain domain. Addressing these challenges the complex service auction extends the body of research on mechanisms for trading combinatorial entities in distributed environments with special focus on sequential compositions of service components in service value networks.

7.4 The Co-Opetition Mechanism

While the technological side to enabling SVNs and related Web service markets and networks, for instance highlighting aspects such as service engineering, service description, or service discovery, enjoys heavy research activity (cp. e.g., Zeng et al. (2003); Cardoso et al. (2004); Barros and Dumas (2006); Küster et al. (2007); Toch et al. (2007); Cardoso et al. (2008); Oberle et al. (2009); Scheithauer et al. (2009)), economic considerations are lagging behind. Yet, a multitude of challenges need to be tackled when entering the market as a mediating party between service customers and service providers. New entrants to the Web service platform market have to demonstrate innovative and competitive business models. In this vein, effective strategies are to sustainably attract market participants from both market sides, customers and providers. Moreover, an attractive service portfolio in terms of variety and quality should be aimed at with the incentives installed through the business model. Therefore, a certain set of objectives is to be fulfilled.

However, besides the seminal contribution of Brandenburger and Nalebuff (1996) that puts *co-opetition* on the agenda as an economic state that is worth to be analyzed, academia provides only little work dedicated to the distinct situation of providers that are related both in *cooperative* and *competitive* terms (Bengtsson and Kock 2000). Coordination in such environments raises the need for schemes and measures that are particularly tailored to co-opetitive value creation.

The objective of this section is to elicit and introduce a novel incentive scheme that is custom-made for SVNs as an organizational form that includes both competition and cooperation amongst service providers. Designing such a scheme, properties to be implemented have to be carefully determined, balanced, and eventually defined. The approach at hand is consult mechanism design in a broader sense following the insight that individuals at the seller side of the SVN cannot merely rely on a selfish strategy as their "fate" in terms of successfully delivering complex services to the customers is inherently and directly connected to other vendors (Iansiti and Levien 2004): service providers have to effectually indicate their willingness to cooperate with distinct other service providers in a complex service offering as a very foundation and requirement for value creation. Therefore, when defining the social choice function of the mechanism, objectives from classic mechanism design, cooperative game theory, and network design come into consideration.

7.4.1 Adaptation of the SVN Formalization

This section uses the SVN-specific notation introduced in Sect. 3.3, which has to be slightly extended to fit the characteristics of the co-opetition mechanism. As previously defined, the graph

$$G := (\{V \cup \{v_s, v_f\}\}, E) \text{ with}$$
$$E := \{e_{i^{\tilde{k}} j^{\tilde{k}+1}}\}, \forall \tilde{k} = 0, \ldots, k$$

represents the fully intermeshed network including source and sink and with all theoretically allowed links between services. However, the platform operator may not be able to "force" service providers to establish links to each and every other eligible service in the network. Therefore, in practice, not all of the possible links between services are necessarily "activated." Let $V_m \subseteq V$ be an arbitrary subset of V and $E(V_m) := \{e_{ij} \in E | v_i, v_j \in V_m \cup \{v_s\}\}$ the set of all associated and reasonable edges. Thus, edges are only included if they form the incoming and outgoing edge of at least one node.[15] In this connection, let $E(v_j) \subset E(V_m)$ denote the set of incoming links that are reasonably associated to a service v_j within V_m (that is, they need to be part of $E(V_m)$).

With respect to the power ratio that is based on the Shapley value (Shapley 1953), as it will be introduced in Sect. 7.4.6, the formalization of complex services needs to be extended. In the following, $F_l := (W_l, E(W_l))$ with $W_l \in \left\{W \subseteq V | \forall k \in \{1, \ldots, \tilde{k}\} \; \exists_1 v_j \in W : v_j \in Y^k\right\}$ shall define a complex service as one element of the set $F := \{F_1, \ldots, F_l, \ldots, F_t\}$ of all t complex services available. Importantly, v_s and v_f are formally excluded from paths due to them not being a service. This definition (1) still fits the requirement that a complex service incorporates exactly one service out of each candidate pool and (2) can be processed by the transfer function (cp. Sect. 7.4.6).

For technical reasons, a reduced graph $\mathcal{G} := (V, E(V))$ is also defined in which v_s, v_f, and $\{e_{if} | v_i \in V^k\}$ are cut off. Further, let $\mathfrak{G}$ denote the set of all possible reduced graphs $\mathcal{G}$.

[15] This is true except for the links of services in $Y^{\tilde{k}}$. In this case, only an incoming node is required.

7.4.2 *Networked Mechanism Design and Requirements in Co-Opetitive Environments*

7.4.2.1 Networked Mechanism Design

Competition in Web service markets is already fierce and will further rise as the SVN concept moves to mainstream. As argued above, effective incentive schemes are required to pull participants onto the offered platform and to make sure that they remain there. An SVN operator has to deliberately cogitate about this issue when launching a new platform. Besides making a profit in the medium term, which short-term goals should be pursued to get a business up and running? As introduced in Sect. 7.1, it is all about *incentive engineering* – and this is exactly what mechanism design is ideally utilized for. Yet, mechanism design in its classic form may be too narrow. This challenge has been notedly illustrated in approaches that waive the classic mechanism design requirement of incentive compatibility, and therefore, also sacrifice allocative efficiency, such as second-best mechanism design (Parkes et al. 2001) and computational mechanism design (Nisan and Ronen 2001).

Yet, in real-world applications and from an economic standpoint, it may not be the ultimate goal to achieve a truthful revelation of the agents' types "at any costs." That way, allocative efficiency is also waived (cp. Sect. 7.1. Sacrificing IC and AE in favor of other properties can be reasonable in order to obtain a "good (enough)" result.

Building thereupon, the *networked mechanism design (NMD) perspective* is introduced in this section. Unlike second-best mechanism design or non-incentive compatible mechanism design, not only classically applied desiderata are sacrificed in order to approximate other objectives. Rather, NMD waives certain classic properties in order to achieve alternative objectives that are closely linked to the respective networked scenario. In case of the application scenario at hand, the NMD approach includes requirements suggested and imposed by networked scenarios in which agents need to co-operate to create value, yet simultaneously competing for the revenues that are collected through the joint value creation activity. This area of conflict is oftentimes inherent to problem sets in networked scenarios – such as incentive engineering in SVNs. The herein included additional desired properties are not in the focus of classic mechanism design.

Akin to second-best mechanism design, classic mechanism design desiderata are likely to fall prey to the newly arising network-related in NMD. The latter do not necessarily have to be a direct consequence of impossibility theorems that limit the design space of mechanisms – as it is the case in second-best mechanism design – but may also be owed to other, prioritized objectives of the desired social choice.

7.4.2.2 Environmental Analysis of SVNs and Consequential Requirements

The following paragraphs briefly discuss the requirements of a mechanism that is capable of supporting the inherent duality of cooperation and competition of service providers in SVNs. Henceforth, the targeted mechanism shall be referred to as "co-opetition mechanism."

The objectives pursued by the co-opetition mechanism are motivated from the platform operator's view in the launch phase of an SVN. In this vein, the main goal is to define incentives that activate network effects and open out into positive feedback loops. That way, self-reinforcing business activity can be created: The more service providers contributing their Web services to the SVN, the higher turns the value of the complex service portfolio that is experienced by the customer. This makes the network more attractive for new vendors, in turn drawing in more customers (Arthur 1996; Shapiro and Varian 1999; Kelly 1999). Consequently, a profit maximization platform operator side shall not be a primary aim at this stage of business and is, therefore, not pursued by the co-opetition mechanism. Supporting and activating a preferably rapid and healthy network development shall be the overall objective and focal point which is reflected in the first two requirements. Thereby, other requirements arise as laid out in Requirements 3–8.

Requirement 1 (Network growth) *The co-opetition mechanism shall incentivize service providers to join the network. That is, it is to be able to foster network growth.*

Requirement 2 (Readiness to deliver) *The co-opetition mechanism shall incentivize service providers to continuously and constantly keep ready their services in the SVN. In other words, service providers shall be rewarded for their constant ability to offer services.*

Requirement 1 aims at attracting service providers that have previously not been a part of the SVN. Requirement 2 is targeted on service providers that are already a part of the SVN, yet may belong to the set of providers that are not regularly allocated, but still constitute the candidate pool and foster the network's diversity. Hence, Requirements 1 and 2 are interdependent: incentives for being ready to deliver may and most probably will influence a service provider's very decision to initially join the network and remain there.

If such readiness shall be accomplished by granting monetary means, payments to a larger set than the allocated service providers in the SVN must be made. It is, therefore, essential to design a fair distribution logic. On the one hand, service providers that make a larger contribution to the network's overall value must receive a larger share of the payment. On the other hand, the distribution logic must be conceived as generally "evenhanded" by all participants to be generally accepted.

Requirement 3 (Fairness) *The co-opetition mechanism shall distribute payoffs amongst service providers as demanded by Requirement 2 in a fair manner.*

Interconnectedness or integration of services is another highly important issue in SVNs since complementarity can only be fully leveraged if the interplay of services is guaranteed in the first place (Economides 1996). As discussed in Sect. 3.3, links e_{ij} in the SVN indicate that the services v_i and v_j are composable. Generally, whether two services are connected depends on two factors: functional and strategic criteria. Since functional interoperability is given due to the requirements imposed by the platform (cp. Sect. 3.2), the decision on being linked to other services or not mainly depends on strategic considerations. In SVNs, links can be formed by individual decision since a service provider cannot influence which other vendors use the output of its service.

Without suitable incentives for link formation, cooperation between service providers that offer complementary functionality is impaired: Customers are expected to prefer purchasing services in networks yielding alternative, substitute offerings such that other providers can dynamically pitch in if an allocated service goes out of business. This is particularly true for critical business applications. Furthermore, such a configuration dramatically decreases customer lock-in. In addition, the promotion of alternative paths through the SVN leads to a more balanced network without single providers that hold monopolistic positions. In balanced networks, the platform operator's dependency on "powerful" service providers, which are able to impose pressure by bulling the market or by threatening the network with termination of membership, can be dramatically mitigated.

Requirement 4 (High degree of interconnectedness) *The co-opetition mechanism shall account for a high degree of interconnectedness among services in the SVN.*

The network design related objectives notwithstanding, the co-opetition mechanism still works within the boundaries of mechanism design. According to Parkes and Kalagnanam (2002b), sustainable mechanisms are at least required to be budget balanced and individually rational. Thus, budget balance and individual rationality constitute the minimal set of classic mechanism design desiderata to be met by the co-opetition mechanism (cp. also Sect. 7.1).

Eventually, the co-opetition mechanism must be able to handle service composition and quality of service characteristics as vital aspects of SVNs (cp. Sect. 3.3). With respect to service composition, it is not sufficient to simply process service bundles without paying attention to the sequence of the elements within the process (Adams and Yellen 1976; Schmalensee 1984; Bakos and Brynjolfsson 1999). As shown in Sect. 3.3, in SVNs one must also account for the services' order in the complex service. Value is only generated if the service modules are in the correct order. Secondly, the multiattributive nature of services needs to be considered since also attributes other than the price play a vital role when allocating a complex service (cp. Sect. 7.2.1).

To complete the list of requirements, the last four requirements explained above are itemized below.

Requirement 5 (Budget balance)

Requirement 6 (Individual rationality)

Requirement 7 (Sequence-sensitive service composition)

Requirement 8 (Quality of service sensitivity)

7.4.3 Service Request and Service Offer

During auction conduction, information needs to be exchanged between the involved parties. The consumer's request is not only represented by the service functionality demanded, but also by the service configuration $\mathscr{A}_{F_l}$ of the complex service $F_l \in F$ (cp. Sects. 3.3 and 7.2.2).

At first, the service customer defines the both functionality demanded and the service attributes of interest to be included. This can be done, for instance, by choosing from a range of possible attributes to be factored into the complex service allocation via an appropriate interface (cp., e.g.,(Bichler et al. 1999)). The comparability of attribute values from different attribute types is ensured by introducing a mapping function Ψ to normalize the attribute values $\mathscr{A}_{F_l}^m$ to an interval [0; 1]. Such a normalization is required to assess the degree of correspondence of F_l to the service consumer's request. The service customer specifies lower and upper boundaries $\Gamma = \left((\gamma_B^1, \gamma_T^1), \ldots, (\gamma_B^m, \gamma_T^m), \ldots, ((\gamma_B^{\tilde{m}}, \gamma_T^{\tilde{m}})\right)$ for each attribute type m. γ_B^m denotes the value of attribute m that results in a zero valuation whereas γ_T^m represents the value of attribute m that yields a maximum fit of 1. If $\mathscr{A}_{F_l}^m \geq \gamma_T^m$, then $\Psi(\mathscr{A}_{F_l}^m) = 1$. Vice versa, $\Psi(\mathscr{A}_{F_l}^m) = 0$ if $\mathscr{A}_{F_l}^m \leq \gamma_B^m$. $\Psi(\mathscr{A}_{F_l}^m)$ is typically assumed to be linear between the upper and lower boundary, however, could also take any other kind of coherency. Importantly, the valuation takes place on a complex service level, i.e., the customer is not interested in the performance of a single service, but in the QoS of the complex service requested.

Further, the customer-specific weighting $\Lambda = (\lambda_1, \ldots, \lambda_m, \ldots, \lambda_{\tilde{m}})$, $\sum_{m=1}^{\tilde{m}} \lambda_m = 1$ is reported which defines the service requester's preferences for each attribute type. This coherency is depicted in the requester's scoring function $\mathscr{Q}$ which includes all non-monetary dimensions of the service and maps them onto a single value (cp. Sect. 7.2.2).

The customer's maximum willingness to pay for a perfect complex service α is the last piece missing in order to fully describe the offered services' fit to the customer's preferences. α is the reservation price for a complex service yielding a score of 1. Altogether, α, the defined weighings Λ, and the customer-specific lower and upper boundaries Γ for each service attribute type constitute the customer type $\theta \in \Theta$ with $\theta = (\alpha, \Gamma, \Lambda)$.

Definition 7.4 (Multiattribute service request). A multiattribute service request for a complex service is a vector defined as follows:

$$SR := (Y^{SC}, \underbrace{\alpha, \Gamma, \Lambda}_{\theta}) \tag{7.8}$$

Y^{SC} represents the desired functionality of the service stated in a suitable format specified by the platform. The maximum customer's willingness to pay for an optimal service configuration is denoted by α. Γ stands for the lower and upper boundary for each attribute type. Finally, Λ represents the requester's preferences for the relevant service attributes.

For simplicity, assume that the service customer states the requested functionality Y^{SC} in the form of the previously introduced candidate pools $(Y^1, \dots, Y^k, \dots, Y^{\tilde{k}})$, with the order of the elements in Y^{SC} indicating the order of the services included in the complex service demanded. Of course, the format specified by the platform can be different, yet, is then to be mapped to the candidate pool logic introduced in Sect. 3.3. α multiplied by $\mathscr{S}(\cdot)$ defines the substitution rate between a complex services' configuration and the requester's preference.

As soon as the requested functionality is submitted to the platform intermediary, potential services belonging to the respective candidate pools are contacted. As a response, service providers willing to participate in the bidding submit their service offers to the intermediary.[16] Service offers consist of a service configuration $\mathscr{A}_j$ and a price bid submitted for a service v_j as successor of v_i. In other words, service providers bid to be included in the complex service dependent on the respective preceding service v_i.

Definition 7.5 (Multiattribute service offer). A multiattribute service offer submitted by a service provider n_h with respect to one of its services v_j, $\bar{\sigma}(v_j) = n_h$, consists of an offered functionality Y^k and a bid $b_{ij} \in \mathscr{B}$ with $b_{ij} = (p_{ij}(e_{ij}), \mathscr{A}_j)$, $v_j \in Y^k, v_i \in Y^{k-1}, k \in \{1, \dots, \tilde{k}\}$:

$$SO_{ij} := (Y^k, \underbrace{p_{ij}(e_{ij}), \mathscr{A}_j}_{b_{ij}}) \tag{7.9}$$

A service provider n_h can submit more than one service offer: both interoperability with more than one service predecessor and the ownership of more than one service that fits the service request result in a set of service offers $SO_h = \{SO_{ij} | \bar{\sigma}(v_j) = n_h,\ v_j \in Y^k,\ v_i \in Y^{k-1},\ k \in \{1, \dots, \tilde{k}\}\}$. The set of all service offers submitted (that lead to the customer-specific SVN G including each of the requested functionalities Y^{SC}) shall be denoted as SO^G.

Example 7.2 (Bidding Language and Scoring Function). The following example illustrates how different attribute types are aggregated along a path of composed service offers in SVNs. It further shows how the set of weights Λ and the upper and lower boundaries Γ are processed in the calculation.

Consider a mid-size company named uServ which seeks to purchase a "service request and order management service" that supports its complex customer relationship business process. The company therefore approaches an SVN, inquiring a solution that (1) handles service requests and orders in a first step (Y^1) and (2) afterwards confirms the orders (Y^2). Further assume that, besides this very functionality Y^{SC}, uServ is mostly concerned with the availability (av) of the service and its throughput (tp), valuing both attributes equally ($\Lambda = \{0.5, 0.5\}$). Additionally,

[16] A more detailed description of the information exchange taking place between the involved parties can be found in Sect. 7.4.4.

uServ defines upper and lower boundaries for av and tp in a certain unit, in this example, assume $\Gamma = ((0.98, 1), (75, 125))$, with γ_1 being measures in percentage, i.e., 0.98 equals 98% availability, and γ_2 being measured in Megabits-per-second (Mbit/s). As uServ is willing to pay $\alpha = 100$ Euros for a defined amount of usages, uServ's type assembles as $\theta = (100, ((0.98, 1), (75, 125)), \{0.5, 0.5\})$, leading to $SR = ((Y^1, Y^2), \theta)$.

The platform operator processes uServ's service request as illustrated in Fig. 7.7, resulting in five service offers SO^G. Exemplarily, service provider n_3's set of service offers SO_3 assembles as $SO_3 = (SO_{13}, SO_{23})$ with $SO_{13} = (Y^2, 20, \{0.998, 110\})$ and $SO_{23} = (Y^2, 33, \{0.998, 110\})$. For the other service offers, please refer to the graph G depicted in Fig. 7.5.

G yields three instantiations: service v_1 combined with service v_3 as the first possibility, service v_2 and service v_3 as the second alternative, and finally a composition of service v_2 and service v_4. So, the set of paths F is defined by $F = (F_1, F_2, F_3)$ with $F_1 = (\{v_1, v_3\}, \{e_{s1}, e_{13}\})$, $F_2 = (\{v_2, v_3\}, \{e_{s2}, e_{23}\})$, and $F_3 = (\{v_2, v_4\}, \{e_{s2}, e_{24}\})$.

As stated earlier and illustrated in Fig. 7.6, it is assumed that the mapping function is linear between the upper and lower boundary such that $\Psi(\mathscr{A}_{F_l}^{av})$ and $\Psi(\mathscr{A}_{F_l}^{tp})$ are defined as follows for each complex service $F_l \in F$.

$$\begin{aligned} \Psi(\mathscr{A}_{F_l}^{av}) &= \frac{1}{1 - 0.98} \cdot (\mathscr{A}_{F_l}^{av} - 0.98) \\ \Psi(\mathscr{A}_{F_l}^{tp}) &= \frac{1}{125 - 75} \cdot (\mathscr{A}_{F_l}^{tp} - 75) \end{aligned} \tag{7.10}$$

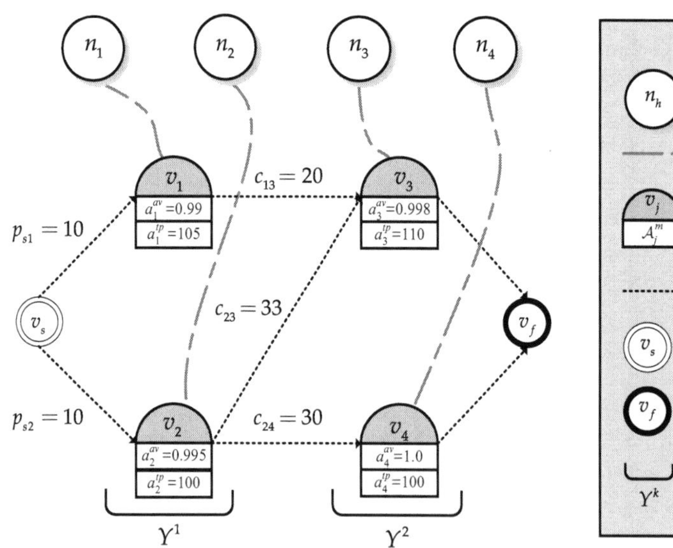

Fig. 7.5 Numerical example of a service value network

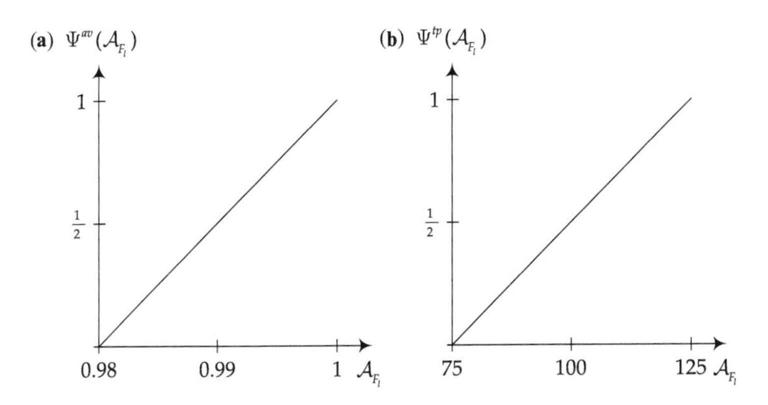

Fig. 7.6 Mapping functions for different attribute types (**a**) uServ's mapping function for different levels of availability (*av*), (**b**) uServ's mapping function for different levels of throughput (*tp*)

Equation (7.10) yields the normalized service configuration. Incorporating Λ, the score $\mathcal{S}(\mathcal{A}_{F_l})$ can finally be computed, which peaks for F_3 (cp. (7.11)).

$$\begin{aligned}\Psi(\mathcal{A}_{F_1}) &= \{0.40, 0.60\} \rightarrow \mathcal{S}(\mathcal{A}_{F_1}) = 0.5 \cdot 0.40 + 0.5 \cdot 0.60 = 0.50\\ \Psi(\mathcal{A}_{F_2}) &= \{0.65, 0.50\} \rightarrow \mathcal{S}(\mathcal{A}_{F_2}) = 0.5 \cdot 0.65 + 0.5 \cdot 0.50 = 0.58\\ \Psi(\mathcal{A}_{F_3}) &= \{0.50, 0.80\} \rightarrow \mathcal{S}(\mathcal{A}_{F_3}) = 0.5 \cdot 0.50 + 0.5 \cdot 0.80 = 0.65\end{aligned} \quad (7.11)$$

7.4.4 Auction Process

In this section, the auction process is shown in more detail. The service customer initiates the formation of a specific SVN while the service providers, formerly registered in the pool of available vendors, bid to be included in a customer-driven SVN. These two sides are brought together to conduct business by the platform operator. The latter takes over the sub-roles of an auctioneer, mechanism operator, and service coordinator.

According to Nisan and Ronen (2001), two basic auction phases can be distinguished – the *declaration phase* and the *execution phase*. In the declaration phase, the necessary information to be exchanged among the participants is gathered. These information objects represent the participants' types which are reported in a direct fashion to the intermediary. To be more precise, the service request submitted by the service customer is processed by the platform operator which reasons about potentially fitting services from the pool of registered services, for instance, relying on semantically supported service discovery (Oberle et al. 2009). Having defined the set of potential services to participate in the mechanism phase, the service intermediary is able to plan and form the actual topology of the SVNs and thereupon sends out a call for bids including additional information on preceding services in order to facilitate context-dependent bidding. This step concludes the preparational activities of the declaration phase and lays the basis for the actual matching – the execution phase.

The execution phase is initiated by the multiattribute service offer submitted by the owners of candidate services which accepted the call for participation. These service offers along with the customer's service request serve as an input for the co-opetition mechanism. An automated search for the optimal path through the network according to the allocation function introduced in Sect. 7.4.6.2 is followed by the calculation of the transfers among network participants by means of the transfer function (cp. Sect. 7.4.6.3). After solving the coordination problem, participants are notified and the actual service delivery (as defined in Sect. 2.1.1) can be launched. During service delivery (which is cautiously put aside in the process model), monitoring techniques can be applied such that the platform operator can cross-check the actual outcome and the reported types of the allocated service providers (Sahai et al. 2002) (cp. also Assumption 7.3 in Sect. 7.4.5).

The auction process is depicted in Fig. 7.7.

7.4.5 Mechanism Assumptions

Before discussing the co-opetition mechanism's implementation, its basic underlying assumptions are stated. First, the co-opetition mechanism shall be centralized. That is, the entire communication is directed via the mechanism which connects all agents and is understood as a self-contained (impartial) entity (Steimle 2008).

Assumption 7.1 (Centrality of the mechanism) *The service customer and all participating service providers are directly connected to the co-opetition mechanism via a fault- and tap-proof communication channel.*

Assumption 7.2 is located at the service provider-side. Since the distribution logic presented in Sect. 7.4.6 grants payoffs to more than the allocated service providers, free-riding becomes an issue. Service providers could publish "dummy services" which are merely designed to skim payments. Therefore, it is required that the services registered with the platform undergo a verification process before they are eligible for an inclusion in customer-specific SVNs.

Assumption 7.2 (No dummy services) *Service provider are not able publish dummy or mock services that do not exhibit the indicated functionality.*

Finally, two general assumptions on the transactions themselves are made. First, consider quality of service. Technically, an agreement between service provider and service consumer about the quality to be delivered must be founded on a legal basis, which is done by specifying a service level agreement (SLA). An SLA is a contract that defines mutual understandings and expectations of a service between service provider and service consumer (Jin et al. 2002). As service providers bid their service configuration which is aggregated over the service attributes' values by each service included in a complex service, the resulting QoS denotes the predefined

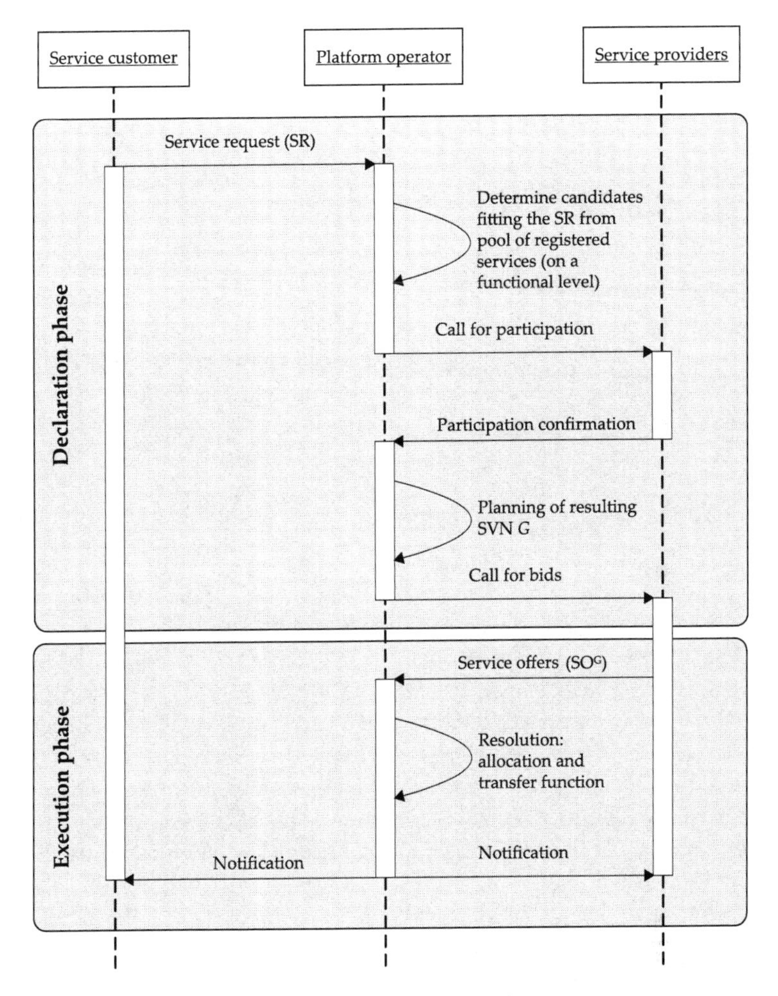

Fig. 7.7 Auction process model: Declaration and execution phase

goal to be fulfilled – i.e., the service level agreed on. The co-opetition mechanism does not include an explicit service level enforcement term. It is, therefore, required that the platform operator can monitor the delivery of a complex service and is able to effectively penalize mal- or non-performance with respect to the SLAs previously agreed on. That is, service providers that register services with the platform commit themselves to offering the services at the quoted price and quality and can thus legally be "forced" to deliver the stated service configuration. For instance, the monitoring concept applied in TEXO shall be capable of backtracking the responsible services if SLAs are not met (cp. Sect. 3.4.3).

Assumption 7.3 (Service level enforcement) *Quoted service configurations can effectively be monitored and penalized in case of mal- or non-performance.*

It is further assumed that the volume of trade is not restricted. Other than in revenue management in which consumer behavior is anticipated in order to maximize profits subject to limited and perishable resources (Talluri and van Ryzin 2004), capacity constraints are not in the focus of the co-opetition mechanism and are deliberately disregarded.

Assumption 7.4 (Unlimited service capacity) *The volume of trade is unrestricted in the co-opetition mechanism.*

If capacity constraints come into play, the reservation of resources in advance becomes an important topic. If the existence of peak times must be rated a serious scenario, there is neither guarantee about offered prices nor the availability of a service. In Chap. 8, this issue is discussed within the scope of dynamic Web services markets.

7.4.6 Mechanism Implementation

In this section, a mechanism is introduced which is to reward service providers based on their contribution made to the network's infrastructure (cp. requirements of the social choice as discussed in Sect. 7.4.2).

The co-opetition mechanism, that will be presented in the remainder of this section, (1) allocates the service modules that together embody the winning complex service and (2) distributes payoffs in a way that is in line with the social choice. The co-opetition mechanism is capable of taking account of the complex service's sequence (cp. Requirement 7) and incorporating quality of service characteristics (cp. Requirement 8) as it is based on the formalization of SVNs (cp. Sect. 3.3).

Recall from Chap. 3 it is a specific customer service request that forms an SVN. Therefore, in the first place, every complex service in the SVN is potentially able to create value. Even if non-allocated, a service has the potential to enrich the platform's variety and thus can contribute to the network. The essence of the co-opetition mechanism is an additionally distributed amount of money Δ according to the so-called *power ratio* .

7.4.6.1 Introducing the Power Ratio

As mentioned above, the co-opetition mechanism's central concept is to compensate not only the set of allocated service providers, but all available service providers that are able to fulfill a specific customer request. This logic shall reflect the network design objectives erected in Requirements 1–4. As mentioned above, in the remainder of this section, the SVN-specific notation introduced in Sect. 3.3 will be consulted. Let further Δ denote a monetary surplus tendered in order to fund the above-stated compensation. δ_j indicates v_j's share in Δ. In general, such a payment t_j granted to each service $v_j \in V$ can be assembled as follows:

$$t_j := \begin{cases} p_{ij} + \delta_j, & \text{if } v_j \text{ is allocated via its link } e_{ij} \\ \delta_j, & \text{if } v_j \text{ is not allocated} \end{cases} \tag{7.12}$$

Consequently, the payoff T_h for a service provider n_h that owns several service offers v_j in the SVN assembles as follows:

$$T_h := \begin{cases} \sum\limits_{\substack{e_{ij} | e_{ij} \text{ allocated}, \\ v_j \in \sigma(n_h)}} p_{ij} + \sum\limits_{v_j | v_j \in \sigma(n_h)} \delta_j, & \text{if at least one } e_{ij} \text{ with } v_j \in \sigma(n_h) \text{ is allocated} \\ \sum\limits_{v_j | v_j \in \sigma(n_h)} \delta_j & \text{if none of } n_h\text{'s services are allocated} \end{cases} \tag{7.13}$$

Example 7.3 (Payoff Distribution). Consider a complex service F_1 that creates a value of x, howsoever calculated. Now assume that there is a different path F_2 exhibiting a value of $x - \varepsilon$. If F_1 was to go out of business, choosing F_2 would only yield a loss of ε. Therefore, the full payoff should not be ascribed to F_1, but also consider any other complex service F_l that has a positive value.

Example 7.3 shows that the idea of the payment introduced in (7.12) and (7.13) directly takes on Requirement 2. By not only rewarding service owners whose services are actually allocated in a specific service composition, but also potential creators of value, service providers that keep ready their resources are granted a percentage of the overall surplus distributed. However, Requirement 1 is also addressed, albeit not as directly as readiness, which can be explained by a reduction of risk in the highly agile SVN context. Service providers are likely to face initial and specific investment costs when designing or adapting services that meet the requirements imposed by the SVN as, for instance, is evident in salesforce.com's AppExchange or in TEXO (cp. Sect. 3.4). Such investments have to be made prior to any transaction, enhancing or facilitating the value of the trade within the platform, but being of considerable less value outside of the platform (Edlin and Reichelstein 1996; Gandal 2002). These specific investments might prompt sellers not to join an SVN since future revenues and transactions are too uncertain compared to the initial investments (Rogerson 1992). Knowing that there will be a recurring payment even if one's service is not regularly allocated might lower the entry barrier for service providers, somewhat providing security through reconciliation of interests in the *initial phase* of the SVN. In order to boost network growth and foster increasing returns, it is inevitable to attract a critical mass of participants. Speaking of service providers, these participants do not necessarily have to be the most competitive ones – as long as the mass of vendors attracted make sure that a sufficiently large number of customers enter the platform. If the SVN is successful in attracting a "good" share of potential customers, previously non-attracted providers might also be impelled to join the SVN (Shapiro and Varian 1999).

By rewarding all potential creators of value, not only the ones that are actually allocated, service providers are not only likely to join the SVN, but also remain there as their readiness to deliver is explicitly rewarded (cp. Requirement 2). These payoffs as sketched in (7.12) shall be realized by the *power ratio (PR)*. The power ratio will be based upon Shapley-style calculus (Shapley 1953).[17] Transferring the logic of the Shapley value to SVNs, the PR can, in short, be interpreted as the average power or significance of a service $v_j \in V$ relative to G. Shifting the traditional notion of purely allocation-based payoffs towards a redistribution among all vendors that are able to provide value for the network is a radical step and necessitates fair distribution (cp. Requirement 3). Such fairness is an inherent property of the Shapley value. Yet, the market would be threatened by adverse selection if the power ratio-based payments were not aligned with the competitive environment. By valuing *marginal contributions*, the Shapley value exhibits appropriate characteristics. Transferred to the objective of SVNs, namely providing customers with complex service offerings, the PR needs to quantify a service provider's marginal contribution at the complex service level.

7.4.6.2 Allocation Function

In this first step of the co-opetition mechanism, the auctioneer solves the winner determination problem. That is, a path $F^* \in F$ is to be found that is in line with the desired outcome.

Let again $\Delta = \sum_{v_j \in V} \delta_j$ be the monetary surplus that is distributed to the set of service providers via the power ratio. Building upon the previous section (in particular, (7.12)), the aggregated surplus $\mathcal{U}^{SP}_{F^*}$ of all service providers owning services in the SVN assembles as the sum of (1) the utility $\mathcal{U}^{SO}_{F^*,o}$ of service offers included in F^* and (2) the utility $\mathcal{U}^{SO}_{F^*,\bar{o}}$ of services that are not a part of F^*:

$$\mathcal{U}^{SP}_{F^*} = \mathcal{U}^{SO}_{F^*,o} + \mathcal{U}^{SO}_{F^*,\bar{o}} = \sum_{\substack{v_j \in W^*, \\ e_{ij} \in E(W^*)}} \left(p_{ij} - c_{ij} + \delta_j\right) + \sum_{v_j \notin W^*} \delta_j \tag{7.14}$$

Let $\mathcal{P}_{F^*} = \sum_{e_{ij} \in E(W^*)} p_{ij}$ represent the sum of the price bids submitted by the service providers for the inclusion of their component services into the allocated complex service F^*. The total payment is collected from the customer which is assumed to be willing to carry the power ratio-based surplus. From a practical point of view, this is a realistic scenario: the SVN enables the customer to purchase a tailored complex service according to its multiattribute service request. In addition, the SVN offers dramatically reduced lock-ins, variety, and resilience due to the potential multitude of possible complex service instances.

[17] For more information on the Shapley value, the interested reader is pointed to the original source (Shapley 1953) or basic books on microeconomics such as Mas-Colell et al. (1995). The concept of the Shapley value has been transferred to networks by Myerson (1977) and Jackson (2005).

Thus, subject to balance according to Requirement 5, the following funds can be distributed: If $\mathscr{P}_{F*}$ is reserved to compensate owners of allocated services for their costs, monetary resources amounting to $\tilde{\mathscr{U}}_{F*}^{SC}$ are available for a power ratio-based distribution Δ

$$\Delta = \sum_{\substack{v_j \in W^*, \\ e_{ij} \in E(W^*)}} \delta_j + \sum_{v_j \notin W^*} \delta_j = \alpha \cdot \mathscr{S}(\mathscr{A}_{F^*}) - \sum_{e_{ij} \in E(W^*)} p_{ij} \tag{7.15}$$

Therefore, $\tilde{\mathscr{U}}_{F_l}^{SC}$ can be interpreted as the value or utility $\mathscr{U}_{F_l} = \alpha \cdot \mathscr{S}\left(\mathscr{A}_{F_l}\right) - \sum_{e_{ij} \in E(W_l)} p_{ij}$ complex service F_l creates in the system, that is, in the customer-specific SVN G.

The actual welfare $\mathscr{W}$ is the sum of the customer utility $\mathscr{U}_{F*}^{SC}$ and the aggregated service provider utility $\mathscr{U}_{F*}^{SP}$ given an allocated complex service F^*. Budget balance requires that the platform neither ploughs money into the network nor withdraws any means from it. Thus, the utility of the platform operator $\mathscr{U}_{F*}^{PO}$ does not influence welfare.

$$\mathscr{W} = \underbrace{\alpha \cdot \mathscr{S}(A_{F^*}) - \sum_{e_{ij} \in E(W^*)} p_{ij} - \Delta}_{U_{F*}^{SC}} + \underbrace{\underbrace{\sum_{v_j \in W^*, e_{ij} \in E(W^*)} \left(p_{ij} - c_{ij} + \delta_j\right)}_{U_{F^*,o}^{SO}} + \underbrace{\sum_{v_j \notin W^*} \delta_j}_{U_{F^*,\bar{o}}^{SO}}}_{\mathscr{U}_{F*}^{SP}} \tag{7.16}$$

From (7.15) and (7.16), the following equation follows:

$$\begin{aligned} \mathscr{W} &= \alpha \cdot \mathscr{S}(A_{F^*}) - \sum_{e_{ij} \in E(W^*)} p_{ij} - \Delta + \Delta + \sum_{e_{ij} \in E(W^*)} \left(p_{ij} - c_{ij}\right) \\ &= \alpha \cdot \mathscr{S}(\mathscr{A}_{F^*}) - \sum_{e_{ij} \in E(W^*)} c_{ij} \end{aligned} \tag{7.17}$$

Yet, the mechanism operator does not have access to the allocated service providers' internal costs when collecting the service request and service offers as this is *private information*. Therefore, the platform operator needs to maximize the expected sum of service customer and provider utility $\tilde{\mathscr{W}} = \bar{E}\left(\mathscr{U}_{F*}^{SC} + \mathscr{U}_{F*}^{SP}\right)$, thereby consulting the assumption of equal internal costs and bid prices of the service providers. That is, the platform operator sets $\sum_{e_{ij} \in E(W^*)} c_{ij} = \sum_{e_{ij} \in E(W^*)} p_{ij}$, leading to the following equation:

$$\tilde{\mathscr{W}} = \bar{E}\left(\mathscr{U}_{F*}^{SC} + \mathscr{U}_{F*}^{SP}\right) = \alpha \cdot \mathscr{S}(\mathscr{A}_{F^*}) - \sum_{e_{ij} \in E(W^*)} p_{ij} \tag{7.18}$$

Based on (7.18), the allocation function o can be defined (cp. Definition 7.6). The optimal path F^* is determined by maximizing the *expected overall utility of the mechanism participants* $\tilde{\mathscr{W}}$.

Definition 7.6 (Allocation function). The allocation function o maps the service providers' bids and the service customer's request to a (feasible) complex service $F^* \in F$ which maximizes $\tilde{\mathscr{W}}$:

$$o := \underset{F_l \in F}{\operatorname{argmax}} \left(\alpha \cdot \mathscr{S}(\mathscr{A}_{F_l}) - \sum_{i,j:\, e_{ij} \in E(W_l)} p_{ij} \right) \tag{7.19}$$
$$\text{s.t.} \qquad \mathscr{U}_{F_l} \geq 0 \ \forall \ F_l \in F$$

The constraint $\mathscr{U}_{F_l} \geq 0 \ \forall \ F_l \in F$ is set to guarantee the customer's individual rationality and budget balance at the same time (cp. Requirements 5 and 6): If $\mathscr{U}_{F_l} < 0 \ \forall \ F_l \in F$ and an allocation was yet to be made, either the platform operator would have to plough money into the transaction or $\mathscr{U}_{F^*}^{SC}$ would turn negative.

In the following, Example 7.2 is continued.

Example 7.4 (Allocation of a complex service in the SVN). Let us go back to the exemplary service value network requested by uServ as depicted in Fig. 7.5, which consists of the three service offers F_1, F_2, and F_3. uServ's preferences and upper and lower boundaries shall be the same as in Example 7.2. The overall utility provided by each of the three services can be computed as follows:

$$\begin{aligned} \mathscr{U}_{F_1} &= 100 \cdot 0.5 - (10 + 20) = 20, \\ \mathscr{U}_{F_2} &= 100 \cdot 0.58 - (10 + 33) = 15, \\ \mathscr{U}_{F_3} &= 100 \cdot 0.65 - (10 + 30) = 25 \end{aligned} \tag{7.20}$$

According to (7.19), the mechanism will then allocate $o = F^* = \operatorname{argmax} \{\mathscr{U}_{F_1}, \mathscr{U}_{F_2}, \mathscr{U}_{F_3}\} = F_3$. The underlying overall utility $\mathscr{U}_{F^*} = \mathscr{U}_{F_3} = 25$ of the best path equals the surplus Δ to be distributed via the power ratio (cp. Example 7.7).

7.4.6.3 Transfer Function

As the second part of the mechanism besides o, this section introduces the monetary transfers to be distributed amongst the involved service providers. As above-mentioned, the essence of the payment function $t(\cdot)$ is an element that values the overall contribution of each service to the specific customer-driven SVN – the *power ratio* t^2 – besides a purely allocation-based element t^1. The latter assures that the provider of an allocated service v_j receives its successful bid (cp. (7.21)). The basic idea of the transfer function was discussed in Sect. 7.4.6.1 in more detail.

$$t_j^1 := \begin{cases} p_{ij}, & \text{if } e_{ij} \in E(W^*) \\ 0, & \text{otherwise} \end{cases} \tag{7.21}$$

The power ratio t^2 is the part which is to implement the bulk of the requirements of the social choice function, capturing and monetizing each service provider's contribution to the overall customer-specific SVN. More precisely, t^2 shall implement the network design goals formulated in Requirements 1–4. As the power ratio is based on solution concepts from cooperative game theory, the concept of so-called coalitions will be used and adapted in order to measure each service's marginal contribution to the SVN.

As the application scenario are SVNs, it is not sufficient to consider coalitions as a mere set of agents; importantly, the network structure has to be incorporated (Jackson 2005). Formally, coalitions need to be replaced by cooperations which are defined as follows: With respect to a given SVN, $S := \{S_1, \ldots, S_m, \ldots, S_{|\mathscr{P}(V)|}\}$ with $S_m := (V_m, E(V_m))$ denotes the set of all *theoretically possible internal cooperations*.[18]

Definition 7.7 (Internal cooperations). Given the set of all service offers SO^G that include the links that are present in a customer-specific SVN, the set of internal cooperations S is defined by the power set of V and their respective attached links. Each of the $\mathscr{P}(V) = 2^{|V|}$ elements in S consist of a set V_m of services attached with their actual links $E(V_m)$.

Example 7.5 (Internal cooperations). Again, please consider the SVN G illustrated in Fig. 7.5. There are $|V| = 4$ services included in G, hence 16 internal cooperations are present. These are as follows: $S_1 = \emptyset$, $S_2 = (\{v_1\})$, $S_3 = (\{v_2\})$, $S_4 = (\{v_3\})$, $S_5 = (\{v_4\})$, $S_6 = (\{v_1, v_2\})$, $S_7 = (\{v_1, v_3\}, \{e_{s1}, e_{13}\})$, $S_8 = (\{v_1, v_4\})$, $S_9 = (\{v_2, v_3\}, \{e_{s2}, e_{23}\})$, $S_{10} = (\{v_2, v_4\}, \{e_{s2}, e_{24}\})$, $S_{11} = (\{v_3, v_4\})$, $S_{12} = (\{v_1, v_2, v_3\}, \{e_{s1}, e_{s2}, e_{13}, e_{23}\})$, $S_{13} = (\{v_1, v_2, v_4\}, \{e_{s2}, e_{24}\})$, $S_{14} = (\{v_1, v_3, v_4\}, \{e_{s1}, e_{13}\})$, $S_{15} = (\{v_2, v_3, v_4\}, \{e_{s2}, e_{23}, e_{24}\})$, $S_{16} = \mathscr{G} = (\{v_1, v_2, v_3, v_4\}, \{e_{s1}, e_{s2}, e_{13}, e_{23}, e_{24}\})$.

It is obvious that only cooperations including complete paths are able to generate value. Therefore, the set F of complex services plays a weighty role when assigning a value to a cooperation. To this end, the concept of characteristic functions from cooperative game theory (Shapley 1953) needs to be, in analogy to Jackson (2005), adopted and extended to so-called *value functions* χ which are rich enough to represent both costs and benefits.

First, it is obvious that a cooperation S_m which does not feature a complete path is assigned a zero value. Second, as soon as a cooperation includes more than one path, the path providing the highest value determines the computation of the

[18]Recall from Sect. 7.4.1 that $E(V_m) := \{e_{ij} \in E | v_i, v_j \in V_m \cup \{v_s\}\}$. $E(v_j) \subset E(V_m)$ was defined as the set of incoming links that are reasonably associated to a service v_j within V_m.

value function. Importantly, the requirement of superadditive value or characteristic functions (as they are required by Shapley (1953) and Myerson (1977)) must be weakened. That is, $\chi((V_1 \cup V_2, E(V_1) \cup E(V_2))) < \chi(S_1) + \chi(S_2)$, $S_1, S_2 \in S$ is accepted as long as $\chi((V_1 \cup V_2, E(V_1) \cup E(V_2)))$ is not smaller than its most valuable component $\chi(S_1)$ or $\chi(S_2)$. In generally this implies for each $1 \leq x \leq |\mathscr{P}(V)|$:

$$\chi\left(\left(\bigcup_{m=1}^{x} V_m, \bigcup_{m=1}^{x} E(V_m)\right) \geq \max\left(\chi(S_1), \ldots, \chi(S_x)\right)\right) \tag{7.22}$$

Example 7.6 (Added value in internal cooperations). Consider a random service v_h that is added to a cooperation S_1 with $V_1 \supseteq W_1$, $S_1 \in S$, $F_1 \in F$. Now assume that v_h does not account for an additional complex service in $(V_1 \cup \{v_h\}, E(V_1) \cup E(v_j))$. Therefore, v_h does not provide any additional value to S_1. Yet, as soon as v_h is added to a cooperation S_2 with $V_2 \supseteq W_2$, $S_2 \in S$, $F_2 \in F$, thereby accounting for a cooperation $S_3 = (V_2 \cup \{v_h\}, E(V_2) \cup E(v_h))$ which yields an additional complex service $F_3 \in F$ with $\chi(F_2) < \chi(F_3)$, then $\chi(S_3) = \max\{\chi(F_2), \chi(F_3)\} = \chi(F_3)$.

In line with (7.15), the value function χ of a complex service $F_l \in F$ is the service configuration-adapted willingness to pay of the service customer net of the sum of the submitted internal prices for the edges included in F_l:

$$\chi(F_l) := \mathscr{U}_{F_l} = \alpha \cdot \mathscr{S}\left(\mathscr{A}_{F_l}\right) - \sum_{e_{ij} \in E(W_l)} p_{ij} \tag{7.23}$$

Hence, SR and SO^G are fully incorporated into the value generated by F_l. Following from Definition 7.6, a third requirement for the value function arises that is closely connected to BB and IR. If a cooperation S_m yields one or more complex services with each of them generating a customer utility less than zero, $\chi(S_m) = 0$ holds. Thus, such a cooperation is treated equally by χ than a cooperation which does not include a path at all.

Based on (7.22) and (7.23), the value function $\chi \in$ X can be defined. The set of all possible value functions is denoted by X.

Definition 7.8 (Value functions of cooperations in the SVN). The value function $\chi : S \rightarrow \mathbb{R}$ maps any $S_m \in S$ to the real numbers. It represents the worth of S_m in terms of its value or utility for the SVN. χ is defined on $\mathscr{G}$ and any of its subsets $S_m \in S$, requiring $\chi(\emptyset) = 0$, $\chi(S_m) = 0$ if $V_m \not\supseteq W_l \; \forall F_l \in F$, $\chi(S_m) = 0$ if $\mathscr{U}_{F_l} < 0 \; \forall W_l \subseteq V_m$, $F_l \in F$, and $\chi(S_m) = \max_{W_l \subseteq V_m} \mathscr{U}_{F_l}$ otherwise, as stated in (7.24).

$$\chi(S_m) := \begin{cases} \max\limits_{W_l \subseteq V_m} \mathscr{U}_{F_l}, & \text{if } \exists W_l \subseteq V_m, F_l \in F, S_m \in S \, \wedge \, \mathscr{U}_{F_l} \geq 0 \\ 0, & \text{if } \nexists W_l \subseteq V_m, F_l \in F, S_m \in S \\ 0, & \text{if } \mathscr{U}_{F_l} < 0 \; \forall W_l \subseteq V_m, F_l \in F, S_m \in S \end{cases} \tag{7.24}$$

At this point, the introduction of the power ratio is fully prepared. To determine the PR of each of the $|V|$ services in the SVN $\mathcal{G} \in \mathfrak{G}$, a function $\phi : \mathfrak{G} \times \mathrm{X} \rightarrow \mathbb{R}^n$ with $\phi_j(\mathcal{G}, \chi) \in \mathbb{R}$ is defined for each service $v_j \in V$. Each service that generates a positive value, that is, $\alpha \cdot \mathcal{S}\left(\mathcal{A}_{F_l}\right) > \mathcal{P}_{F_l}$ holds for at least one complex service the respective service is a part of, is considered *vital* in at least one instantiation of the customer request.

Taking into account (7.24) and the consideration of each sub-network $S_m \in S$ of $\mathcal{G}$ into the Shapley value,[19] (7.25) yields the *power ratio* of service v_j.

$$\phi_j(\mathcal{G}, \chi) = \sum_{S_m \in S | v_j \in V_m} \gamma_{S_m} \cdot \left(\chi(S_m) - \chi(S_m^{-j})\right) \tag{7.25}$$

with

$$\gamma_{S_m} = \frac{(|V_m| - 1)! \cdot (|V| - |V_m|)!}{|V|!} \tag{7.26}$$

S_m^{-j} stands for the tuple in which v_j and $E(v_j)$ are removed from S_m, i.e., $S_m^{-j} = (V_m \setminus \{v_j\}, E(V_m) \setminus E(v_j))$. The set of all reasonable links of a service v_j within a cooperation $S_m \in S$ is denoted $E(v_j)$ (cp. Sect. 7.4.1). As soon as v_j is included in a cooperation S_m, $E(v_j)$ is also added.

For all internal cooperations $S_m \in S$ a service v_j can in theory be part of, the rightmost term of (7.25) is positive whenever v_j is pivotal to S_m. Thus, it reflects the marginal contribution of v_j to the considered internal cooperation. This value is weighted by the formation probability γ_{S_m} of the underlying cooperation, thereby consulting the logic introduced by Shapley (1953).[20] Let $\Phi(\mathcal{G}, \chi) := (\phi_1, \ldots, \phi_n)$.

Definition 7.9 (Power ratio-based transfer function (PRTF)). For any $v_j \in V$, the PRTF consists of a directly allocation-dependent element t_j^1 and an element t_j^2 that accounts for the overall network view based on Shapley-style calculus:

$$t_j = t_j^1 + t_j^2 := \begin{cases} p_{ij} + \phi_j, & \text{if } v_j \in W^*, e_{ij} \in E(W^*) \\ \phi_j, & \text{otherwise} \end{cases} \tag{7.27}$$

Equation (7.27) can be transferred to the service provider level and consequently evolves as the sum of all transfers granted to services that are owned by provider $n_h \in N$:

$$T_h := \begin{cases} \sum\limits_{\substack{e_{ij} | e_{ij} \in E(W^*), \\ v_j \in \sigma(n_h)}} p_{ij} + \sum\limits_{e_{ij} | v_j \in \sigma(n_h)} \phi_j, & \text{if } \exists e_{ij} \in E(W^*) \text{ with } v_j \in \sigma(n_h) \\ \sum\limits_{e_{ij} | v_j \in \sigma(n_h)} \phi_j, & \text{otherwise} \end{cases} \tag{7.28}$$

[19]Or its application to networks provided by Myerson (1977) (the "Myerson value"), respectively.

[20]For a detailed explanation and interpretation of the combinatorics in (7.26), please refer to technical literature such as Roth (1988), Mas-Colell et al. (1995), or Winter (2002).

Example 7.7 (Calculation of the PRTF). Again, please recall Examples 7.2 and 7.4. Since $F^* = F_3$, the providers of the involved services v_2 and v_4 are both compensated amounting to their bid prices as set by t^1 (cp. (7.21)). On top, $\Delta = \chi(F_3) = 25$ is available for a distribution via the power ratio. Equations (7.25) and (7.26) result in the payoff vector $\Phi = (3.75, 8.75, 6.25, 6.25)$. Exemplarily, the computation of ϕ_4 is shown in detail. In the underlying SVN, service v_4 can be part of eight internal cooperations (cp. also Example 7.5) as listed below:

$$\begin{aligned}
\phi_4 = {} & \underbrace{\left[\left(\frac{(1-1)!\cdot(4-1)!}{4!}\right)\cdot(0-0)\right]}_{S_5=(\{v_4\},E(V_5))} + 2\underbrace{\left[\left(\frac{(2-1)!\cdot(4-2)!}{4!}\right)\cdot(0-0)\right]}_{\substack{S_8=(\{v_1,v_4\},E(V_8)),\\ S_{11}=(\{v_3,v_4\},E(V_{11}))}} \\
& + \underbrace{\left[\left(\frac{2}{24}\right)\cdot(25-0)\right]}_{S_{10}=(\{v_2,v_4\},E(V_{10}))} + \underbrace{\left[\left(\frac{(3-1)!\cdot(4-3)!}{4!}\right)\cdot(25-0)\right]}_{S_{13}=(\{v_1,v_2,v_4\},E(V_{13}))} \\
& + \underbrace{\left[\left(\frac{2}{24}\right)\cdot(20-20)\right]}_{S_{14}=(\{v_1,v_3,v_4\},E(V_{14}))} + \underbrace{\left[\left(\frac{2}{24}\right)\cdot(25-15)\right]}_{S_{15}=(\{v_2,v_3,v_4\},E(V_{15}))} \\
& + \underbrace{\left[\left(\frac{(4-1)!\cdot(4-4)!}{4!}\right)\cdot(25-20)\right]}_{S_{16}=(\{v_1,v_2,v_3,v_4\},E(V_{16}))=\mathcal{G}} = 6.25
\end{aligned}$$

It is directly conceivable from the equation shown above that service v_4 is pivotal in four internal cooperations. For example, v_4 is pivotal for S_{16}, decreasing its value from $\chi(S_{16}) = 25$ to $\chi(S_{16}^{-4}) = \chi((V_{16} \setminus \{v_4\}, E(V_{16}) \setminus E(v_4))) = 20$ when it is removed. This difference is then weighted by the probability of the cooperation to form. The other services' power ratios are computed analogously.

In sum, applying (7.27), the owners n_2 and n_4 of the allocated services v_2 and v_4 receive their price bid and their PRTF, resulting in $T_2 = t_2 = 10 + 8.75 = 18.75$ Euros and $T_4 = t_4 = 30 + 6.25 = 36.25$ Euros. Non-allocated services (that is, their owners) receive their PRTF share only: Service provider n_1 gets a payment of $T_1 = t_1 = 3.75$ Euros and vendor v_3 receives $T_3 = t_3 = 6.25$ Euros. uServ is charged $p_{s2} + p_{24} + \chi(F_3) = 65$ Euros for the requested complex service.

7.4.7 Summary

Section 7.4 presented the co-opetition mechanism implementation, thereby explaining the concrete allocation and transfer function of the co-opetition mechanism which are designed to fulfill the requirements imposed by SVNs. The power ratio was introduced as the heart of the co-opetition mechanism to install suitable incentives to realize the social choice. The PR, thereby, sets the vision of distributing value in commercial service networks by rewarding all potential value creators, not only the "best" ones.

Theoretic results that directly follow from the mechanism implementation in terms of the social choice are briefly recapitulated below:

- Requirement 2 is directly reflected by the power ratio: service providers are rewarded for "standing by" with their service offer – even if they are not allocated.
- The co-opetition mechanism meets budget balance (cp. Requirement 5) by design. In the co-opetition mechanism, the allocation and transfer rule can be realized without external subsidies. The first component of the transfer function assures that allocated service providers receive their price bids p_{ij} while the service customer compensates for $\sum_{e_{ij} \in E(W^*)} p_{ij}$ (cp. Definition 7.9). Additionally, the power ratio distributes $\Delta = \alpha \cdot \mathscr{S}(\mathscr{A}_{F^*}) - \sum_{e_{ij} \in E(W^*)} p_{ij}$ amongst all services $v_j \in V$. This surplus is also fully distributed by the construction of the PR. As this premium is borne by the service customer, too, additional payments to back up the monetary flows between the participating actors are not required.
- As already stated above, Requirements 7 and 8, which are necessary to ensure the applicability of the mechanism in SVNs, are met through the application of the underlying model (cp. Sect. 3.3) and bidding language (cp. Sect. 7.2.2).

The idea of the co-opetition mechanism builds on "second-best mechanism design" (Parkes et al. 2001) as a variant of classic mechanism design, establishing a notion that a waiver of traditionally pursued objectives such as incentive compatibility and allocative efficiency (cp. Sect. 7.1 can be very useful and appropriate in certain application domains. Thus the co-opetition mechanism may be seen as an application of NMD in which network-related objectives overbalance the desiderata from classic mechanism design.

Evidence for the fulfillment of the remaining requirements stated in Sect. 7.4.2 can be found in Conte et al. (2009), Conte et al. (2010), and Conte (2010).

References

Adams W, Yellen J (1976) Commodity bundling and the burden of monopoly. Q J Econ 90(3): 475–498

Andrieux A, Czajkowski K, Dan A, Keahey K, Ludwig H, Pruyne J, Rofrano J, Tuecke S, Xu M (2004) Web services agreement specification (WS-Agreement). In: Global Grid Forum

Arthur W (1996) Increasing returns and the new world of business. Harv Bus Rev 74:100–111

Asker J, Cantillon E (2008) Properties of scoring auctions. RAND J Econ 39(1):69–85

Bakos Y, Brynjolfsson E (1999) Bundling information goods: Pricing, profits, and efficiency. Manag Sci 45(12):1613–1630

Barberà S, Jackson M (1995) Strategy-proof exchange. Econometrica 63(1):51–87

Barros A, Dumas M (2006) The rise of web service ecosystems. IT Prof 8(5):31–37
Beall S (2003) The role of reverse auctions in strategic sourcing. Tech. rep., CAPS Research
Bengtsson M, Kock S (2000) "coopetition" in business networks – to cooperate and compete simultaneously. Ind Market Manag 29(5):411–426
Bichler M (2001) The future of e-markets: Multidimensional market mechanisms. Cambridge University Press, Cambridge
Bichler M, Kalagnanam J (2006) Software frameworks for advanced procurement. Commun ACM 49(12):104–108
Bichler M, Kaukal M, Segev A (1999) Multi-attribute auctions for electronic procurement. Tech. rep.
Bichler M, Pikovsky A, Setzer T (2009) An analysis of design problems in combinatorial procurement auctions. Bus Inform Syst Eng 1(1):111–117
Blau B (2009) Coordination in service value networks – a mechanism design approach. PhD thesis, Universität Karlsruhe (TH)
Branco F (1997) The design of multidimensional auctions. RAND J Econ 28(1):63–81
Brandenburger A, Nalebuff B (1996) Co-opetition. Doubleday, New York
Cardoso J, Sheth A, Miller J, Arnold J, Kochut K (2004) Quality of service for workflows and web service processes. Web Semant Sci Serv Agents World Wide Web 1(3):281–308
Cardoso J, Voigt K, Winkler M (2008) Service engineering for the internet of services. Enterprise Inform Syst 10:15–27
Chandrashekar T, Narahari Y, Rosa C, Kulkarni D, Tew J, Dayama P (2007) Auction-based mechanisms for electronic procurement. IEEE Trans Autom Sci Eng 4(3):297–321
Che Y, Institute SSR of Wisconsin-Madison U (1993) Design competition through multidimensional auctions. RAND J Econ 24:668–668
Clarke E (1971) Multipart pricing of public goods. Publ Choice 11(1):17–33
Conte T (2010) Value creation through co-opetition in service networks. Dissertation, Karlsruhe Institute for Technology
Conte T, Blau B, Satzger G, van Dinther C, Weinhardt C (2009) Rewarding contribution to service network formation. In: Proceedings of the 1st INFORMS international conference on service science, Hong Kong, paper ID 225
Conte T, Blau B, Xu Y (2010) Competition of service marketplaces – designing growth in service networks. In: Proceedings of the 18th European conference on information systems (ECIS), Pretoria, paper ID 287
Cramton P, Shoham Y, Steinberg R (2006) Combinational auctions. MIT, Cambridge
Dan A, Ludwig H, Pacifici G (2003) Web service differentiation with service level agreements. White Paper, IBM Corporation
Dash R, Jennings N, Parkes D (2003) Computational-mechanism design: A call to arms. IEEE Intell Syst 18(6):40–47
Dijkstra E (1959) A note on two problems in connexion with graphs. Numer Math 1(1):269–271
Economides N (1996) The economics of networks. Int J Ind Organ 14(6):673–699
Edelman B, Ostrovsky M, Schwarz M (2007) Internet advertising and the generalized second-price auction: Selling billions of dollars worth of keywords. Am Econ Rev 97(1):242–259
Edlin A, Reichelstein S (1996) Holdups, standard breach remedies, and optimal investment. Am Econ Rev 86(3):478–501
Engel Y, Wellman MP, Lochner KM (2006) Bid expressiveness and clearing algorithms in multiattribute double auctions. In: Proceedings of the 7th ACM conference on electronic commerce. ACM, New York, pp 110–119
Gandal N (2002) Compatibility, standardization, and network effects: Some policy implications. Oxf Rev Econ Pol 18(1):80–91
Groves T (1973) Incentives in teams. Econometrica 41(4):617–631
Iansiti M, Levien R (2004) Strategy as ecology. Harv Bus Rev 82(3):68–81
Jackson M (2003a) Efficiency and information aggregation in auctions with costly information. Rev Econ Des 8(2):121

Jackson M (2003b) Mechanism theory. In: Derigs U (ed) The encyclopedia of life support systems. EOLSS Publishers, Oxford
Jackson M (2005) Allocation rules for network games. Games Econ Behav 51(1):128–154
Jap S (2002) Online reverse auctions: Issues, themes, and prospects for the future. J Acad Market Sci 30:506–525
Jin L, Machiraju V, Sahai A (2002) Analysis on Service Level Agreement Of Web Services. Technical report, HP Laboratories
Kelly K (1999) New rules for the new economy: 10 ways the network economy is changing everything. Penguin Books, New York
Klemperer P (2004) Auctions: Theory and practice. Princeton University Press, Princeton
Kothari A, Parkes D, Suri S (2005) Approximately-strategyproof and tractable multiunit auctions. Decis Support Syst 39(1):105–121
Küster U, König-Ries B, Stern M, Klein M (2007) Diane: An integrated approach to automated service discovery, matchmaking and composition. In: Proceedings of the 16th international conference on world wide web (WWW), Banff, pp 1033–1042
Lucking-Reiley D (2000) Auctions on the internet: What's being auctioned, and how? J Ind Econ 48(3):227–252
Mas-Colell A, Whinston M, Green J (1995) Microeconomic theory. Oxford University Press, New York
Matthyssens P, Vandenbempt K (1998) Creating competitive advantage in industrial services. J Bus Ind Market 13:339–355
McAfee R (1992) A dominant strategy double auction. J Econ Theor 56(2):434–450
Milgrom P (2004) Putting auction theory to work. Cambridge University Press, Cambridge
Myerson R (1977) Graphs and cooperation in games. Math Oper Res 2(3):225–229
Myerson R (1988) Mechanism design. Discussion Papers 796, Northwestern University, Center for Mathematical Studies in Economics and Management Science.
Myerson R, Satterthwaite M (1983) Efficient mechanisms for bilateral exchange. J Econ Theor 28:265–281
Neumann DG (2004) Market engineering – A structured design process for electronic markets. PhD thesis, Universität Karlsruhe (TH)
Nisan N (2007) Introduction to mechanism design (for computer scientists). In: Nisan N, Roughgarden T, Tardos E, Vazirani V (eds) Algorithmic game theory. Cambridge University Press, New York, pp 209–242
Nisan N, Ronen A (2001) Algorithmic mechanism design. Games Econ Behav 35(1–2):166–196
Oberle D, Bhatti N, Brockmans S, Niemann M, Janiesch C (2009) Countering service information challenges in the internet of services. Bus Inform Syst Eng 1(5):370–390
Papadimitriou C, Steiglitz K (1998) Combinatorial optimization: Algorithms and complexity. Dover Publications, New York
Papazoglou P (2008) Web services: Principles and technologies. Prentice Hall, Upper Saddle River
Parkes D (2001) Iterative combinatorial auctions: Achieving economic and computational efficiency. PhD thesis, University of Pennsylvania
Parkes D, Kalagnanam J (2002a) Iterative multiattribute vickrey auctions. Tech. rep., Harvard University
Parkes D, Kalagnanam J (2002b) Iterative multiattribute vickrey auctions. Tech. rep., Harvard University
Parkes D, Kalagnanam J (2005) Models for iterative multiattribute procurement auctions. Manag Sci 51(3):435–451
Parkes D, Kalagnanam J, Eso M (2001) Achieving budget-balance with vickrey-based payment schemes in combinatorial exchanges. Tech. rep., IBM Research
Pesendorfer W, Swinkels J (2000) Efficiency and information aggregation in auctions. Am Econ Rev 90(3):499–525
Rogerson W (1992) Contractual solutions to the hold-up problem. Rev Econ Stud 59(4):777–793
Ronen A, Lehmann D (2005) Nearly optimal multi-attribute auctions. In: Proceedings of the 6th ACM conference on electronic commerce. ACM, New York, pp 279–285

Roth A (1988) Introduction to the Shapley value. In: Roth A (ed) The Shapley value: Essays in honor of Lloyd S. Shapley. Cambridge University Press, Cambridge, pp 1–27

Rothkopf M, Pekeč A, Harstad R (1998) Computationally manageable combinational auctions. Manag Sci, pp. 1131–1147

Sahai A, Machiraju V, Sayal M, Van Moorsel A, Casati F (2002) Automated SLA Monitoring for Web Services, Proceedings of the 13th IFIP/IEEE International Workshop on Distributed Systems: Operations and Management (DSOM), Montreal, pp. 28–41

Sandholm T, Suri S, Gilpin A, Levine D (2005) CABOB: A fast optimal algorithm for winner determination in combinatorial auctions. Manag Sci 51(3):374–390

Scheithauer G, Voigt K, Bicer V, Heinrich M, Strunk A, Winkler M (2009) Integrated service engineering workbench: Service engineering for digital ecosystems. In: Proceedings of the international conference on management of emergent digital ecosystems (MEDES), Lyon, pp 446–449

Schmalensee R (1984) Gaussian demand and commodity bundling. J Bus 57(1):211–230

Shapiro C, Varian H (1999) Information rules. Harvard Business School Press, Boston

Shapley L (1953) A value for n-person games. In: und AW Tucker HK (ed) Contributions to the theory of games, vol II. Princeton University Press, Princeton

Shoham Y, Leyton-Brown K (2008) Multiagent systems: Algorithmic, game-theoretic, and logical foundations. Cambridge University Press, New York

Smith C (1989) Auctions: The social construction of value. University of California Press, Berkeley

Steimle J (2008) Algorithmic mechanism design: Eine Einführung. Springer, Berlin

Talluri K, van Ryzin G (2004) The theory and practice of revenue management. Springer, Berlin

Toch E, Gal A, Reinhartz-Berger I, Dori D (2007) A semantic approach to approximate service retrieval. ACM Trans Internet Technol (TOIT) 8(1):2:1–2:29

Varian H (2009) Online Ad auctions. Am Econ Rev (99):430–434

Vickrey W (1961) Counterspeculation, auctions, and competitive sealed tenders. J Financ 16(1):8–37

Weinhardt C, Holtmann C, Neumann D (2003) Market engineering. Wirtschaftsinformatik 45(6):635–640

Wellman P (2005) Online marketplaces. In: Singh M (ed) The practical handbook of internet computing. CRC Press, Boca Raton, pp 1–17

Winter E (2002) The Shapley value. In: Aumann R, Hart S (eds) Handbook of game theory with economic applications, handbooks in economics 11, vol 3. North-Holland, Amsterdam, pp 2025–2054

Zeng L, Benatallah B, Dumas M, Kalagnanam J, Sheng QZ (2003) Quality driven web services composition. In: Proceedings of the 12th international conference on World Wide Web. ACM, New York, pp 411–421. doi:http://doi.acm.org/10.1145/775152.775211

Chapter 8
Web Services Advanced Reservation Contracts

8.1 Reservations in Web Service Markets

In Chap. 5, the benefits which dynamic pricing mechanisms hold for providers as well as customers were pointed out. Though already commonly used in practice both market participants are faced with certain risks: Web service providers face the risk of low incomes in times of (unexpected) low utilization and demand, and are thus forced to lower their prices accordingly in order to keep up with their competitors. This may result in considerable losses. On the other hand, Web service customers may run into situations where, despite of a high willingness to pay, cannot fulfill their demand in peak times. While for the provider this dilemma has mainly direct economic consequences only, it can become a more serious issue to the customers than only high costs.

In agreement with Papazoglou (2008) from a customers point of view, Web services can be classified into *replaceable* and *mission-critical* services. Replaceable services can be thought of as services being offered by multiple providers. As the acquisition is thus not dependent on a monopolistic infrastructure, these services and their inherent business functionality can easily be requested from any of those providers, given that the technical interfaces are compliant. Mission-critical Web services on the other hand are more crucial to business processes and even more sensitive to shortages, as they usually are only offered by a few, highly specialized providers. As these services often require a certain level of customization and are not adaptive, in most cases they cannot be replaced easily.

As these services are responsible for the flawless execution of whole business processes the risk of probably not being able to acquire these services, regardless of the price to pay, is not acceptable for most business customers. A similar issue was also already noticed in other related areas, namely Grid and Cloud Computing, where computing resource shortages and their sparse availability lead to similar problems. Solutions where proposed amongst others by Meinl (2008): The basic idea is to complement the dynamic pricing systems by a second market mechanism enabling the advanced reservation of resources and services. In this additional so

C. Weinhardt et al., *Business Aspects of Web Services*,
DOI 10.1007/978-3-642-22447-8_8,

called *contract market* (see also Kleindorfer and Wu (2003)) both Web service provider and customer close reservation contracts where they agree upon the delivery of a specific service at a specific future time spot for a certain price. Thus, this market is the counterpart to the in the previous chapters pointed out Web service spot markets, where customers arrive without any prior notification and accept the current prevalent price as well as the risk of their request not being satisfied. Yet, both markets have their reason to exist: The contract market for all kinds of mission-critical Web services, and the spot market for the replaceable ones.

In the following sections it will be analyzed what kind of Web service contract markets are appropriate and on what assumptions and requirements they rely. Basically, one can distinguish between two models: So called *derivatives markets* which have already been established in the financial end energy sector, and *yield management* techniques which became popular foremost in the airline industry. Independent of the utilized model the main question that remains is how much the provider should charge for offering a reservation contract. This *reservation price* calculation is not trivial, as in dynamic markets the price of the service to be delivered in the future is unknown at the time the reservation contract is closed. It will be shown how the principles of derivative pricing and optimized provider revenues can be transferred and adapted from established areas. Both approaches allow us to abstract from the actual dynamic pricing model used in the spot market and are thus widely applicable.

8.2 Web Service Derivatives

In this section it is assumed that the price for each Web service offered at the spot market is given through the stochastic price process X, that is, for each time spot t the Web service can be acquired without any prior reservation for $X(t)$ only. As it is also assumed that the price processes for individual services are independent from another, w. l. o. g. the focus lies on the price process of one particular Web service. One implication of this assumption is that in this scenario there is a perfectly competitive market, that is, all providers offer the same (or equivalent) Web service at the same price. This already constitutes the main difference to the yield management model presented in Sect. 8.3 where the subject of markets exhibiting characteristics as imperfect competition is discussed. Some materials and ideas in this section are partially adapted from Meinl and Blau (2009).

8.2.1 Derivatives: Definition and Application Areas

Derivatives are nowadays established instruments in financial, commodity and other markets. They are often used as tools with leverage effects that can be designed in a highly flexible manner. Hull (2008) defines a derivative as a *financial instrument whose value depends on (or derives from) the values of other, more basic, underlying variables*. The most common derivatives are the so called *Futures* and *Options* contracts: While a Futures contract obligates the holder to buy or sell an asset at a predetermined delivery price during a specified future time period, an option affords the holder only with the right but *without* any obligation to buy or sell the asset agreed upon in the contract. In a more mathematical framework, Björk (2009) defines a derivative in terms of a *contingent claim* as follows.

A *contingent claim* with *date of maturity* (exercise date) T, also called T-claim, is any stochastic variable $\mathcal{X} \in \mathcal{F}_T^X$. A contingent claim $\mathcal{X}$ is called a *simple claim* if it is of the form

$$\mathcal{X} = \Phi(X(T)). \tag{8.1}$$

The function Φ is called the *contract function*.

The advantage of above mathematical formulation of derivatives is that the contract function Φ serves us as a abstraction of the derivative itself. Thus, for the rest of this chapter, there is no need to explicitly consider all kind of different derivative types, but rather all derivatives can be treated the same way by referring to their contract function only. The contract function specifies the worth of the contingent claim at its exercise date. The terms derivative and contingent claim stem from the fact that these contracts directly depend on the realization of the underlying stochastic price process of the asset Φ is defined upon. In financial markets, these assets are usually stocks or commodities like oil, hogs, cattle or corn. Since then, the notion of derivatives has been extended to include other (also intangible) assets, like electricity. This section proposes the establishment of derivatives on Web services, which share similar characteristics.

Today's most prominent example in the literature of a simple contingent claim is the *European call option*. Being one of the most commonly traded derivatives in financial markets, it gives the owner the right, but not the obligation, to buy a certain number of asset it is defined upon, for a predetermined price at the date of maturity T. Thus, today the option is worth the by the continuously compounded risk free rate discounted difference of the yet unknown stock price at T and the option's exercise price. If this difference is not positive, the option is worth zero. Note that in the here proposed Web service derivative approach only derivatives that have a fixed exercise date are considered, that is, T. Though other kinds of derivatives, that allow for the advance reservation of a Web service during a time period $[T_1, T_2]$, $T_1 < T_2$, rather than one date T, may seem promising and utile, for many of these derivative pricing models there exists no closed form solution. In these cases, pricing, though following the same principles, can be achieved through, for example, simulations.

Today derivatives are mainly used for two purposes: Risk hedging and investment (that is, speculation). As risk hedging instruments they have long been established in financial and commodity markets, where future prices are uncertain. Usage of derivatives foster long-term strategic planning without exposing the financial budget to unnecessary uncertainties. While this holds particularly for commodity markets, trading derivatives for monetary investments take place in both markets: The usage of derivatives to amplify leverage effects in stock trades as well as the virtual trade (i.e., the trade of derivatives based on commodities with no intention to actually acquire these in the end) have become common procedures. Lately, derivatives have also established themselves in other areas, like electricity markets[1] where the daily trade of call and so called swing options has been commonly established since 2002. The exercise of derivatives in these markets is not connected with any delivery of the underlying, but is settled solely in cash.

Derivatives have become common trading objects used for the linked purposes of risk hedging and investment in the most diverse areas. Though there have been recent advances to open up new markets, not all have been successful as could be observed at the prominent example of the Cantor Exchange[2] which planned to offer futures based on the basis of box office receipts. New regulations and legislations trying to prevent the misuse of these originally financial instruments may impede their advance.

Thus, a Web service derivative can be seen as the generalization of a reservation contract to ensure the availability of a Web Service at a fixed price. This allows us to handle reservations in a very flexible manner, including not only the treatment of static reservations (like financial futures) but also all other kind of reservation contracts providing the service customer with the desired degree of risk hedging and flexibility. Though until this day there has not yet been established a market for Web service derivatives, it is possible to look at the principles and rules of their pricing in other (primarily financial) markets and to analyze how they must be adapted in order to be compliant to Web services.

8.2.2 *Principles of Derivative Pricing*

Derivatives in financial and other markets are and should always be priced according to the principles of efficiency by using the market's completeness. Following Hull (2008) and Björk (2009) these crucial principles are as follows.

- *Arbitrage.* Arbitrage denotes the (monetary) advantage exploited when two or more financial instruments are mispriced in relation to another. By exploiting these mispricings, any trader being aware of these, can make a riskless profit that otherwise would not be possible. Therefore, when establishing and pricing new instruments in financial or alike markets, it is the primary aim that no arbitrage opportunities are introduced. This, of course, must also hold for all already traded instruments. However, as Hull (2008) notes, in markets with sufficient liquidity and trading volume (including also the amount of traders) any arbitrage opportunities will be noticed almost immediately by many traders. Therefore, as these traders will try to exploit the mispriced instruments, these will vanish as

[1] http://www.eex.com/.

[2] http://www.cantorexchange.com/.

quickly as they arose, leaving only a very short time window to act. Yet, with today's establishment of electronic markets and automated trading, this issue has aroused interest again.

- *Efficiency.* A market is called efficient if it does not allow for any arbitrage opportunities, that is, all in the market traded financial instruments must be priced accordingly. Therefore, when designing the market rules for trading and pricing, this subject is of crucial importance, also to protect the less informed traders and giving equal chances to all market participants. In case of continuous mispricings the market could be thrown off-balance with severe consequences and would put off traders, resulting in less liquidity and trading volume.
- *Completeness.* While efficiency is one of the principle aims in market design regarding the rules of pricing, completeness is one of the means how efficient pricing can be achieved. In a complete market, all traded objects can be replicated, that is, for every financial instrument there exists a self-financing trading strategy which for every $t \leq T$ replicates the state of this instrument, generally by forming a portfolio of other assets.

Thus, completeness implicates that in such markets (theoretically) the replicated financial instrument itself becomes obsolete. The concept of completeness is mainly a theoretical one used as the basic theory for pricing derivatives. In practice such a self-financing strategy would require to adapt the respective portfolio in continuous time, neglecting all aspects like trading taxes, lags and availability issues. The basic idea is to determine the price of every derivative based on its equivalent portfolio which must not contain the derivative itself. In this way, different derivatives based on the same underlying assets can be priced in a unique way without introducing any arbitrage opportunities. By using the model that implements this idea, this unique price can be determined for any $t < T$, depending on the contract function and the underlying stochastic price process. Using an equivalent portfolio that mimics exactly the state of the derivative at any time spot, essentially renders this claim redundant.

8.2.3 *Web Service Derivative Pricing*

In the previous section, it was depicted according to what principles derivatives in financial markets are priced. It will now be analyzed how these models can be transferred or must be altered to price derivatives defined on Web services. As it was seen in order to (uniquely) price a derivative the only thing one has to do is to set up a self-financing portfolio which mimics the behavior of the respective derivative and is theoretically adapted in continuous time. Generally, this portfolio consists at least of some risk free asset and the underlying itself. While it can be assumed that the risk free asset (e.g., in form of a bank account) poses no problem, that is, it is always available in the actual model, matters are different with the underlying. For financial options one can hold the necessary number of stocks to form a portfolio. From a theoretical point of view it is necessary that the amount of stocks (or other tradeable goods) included in the portfolio are not limited to integers but can be managed in a frictionless manner, that is, in real numbers. As this cannot hold in reality, one must keep in mind that the here presented models are only an approximation of the theoretical optimal (i.e., fair) price. This point is further substantiated by the fact that the models also assume, that any change in the portfolio can be done without an time delay. Nevertheless the models are accepted and applied in practice. As mentioned, it is possible to buy any commodities and stocks in order to make them part of the portfolio derivative, at least in theory. But what about Web services?

As was pointed out in Chap. 4 Web services exhibit certain special characteristics that distinguish them from ordinary goods, like intangibility and non-storability. As a direct consequence of these characteristics, Web services cannot be stored, resold, or actively traded any further, once they have been acquired directly on the spot market without using any reservation in advance system. Therefore, from the perspective of a Web service derivative portfolio, it is not feasible to include the Web service itself into this portfolio, thus, preventing to form a replicating portfolio as is done in the models of complete markets. While in this case the price of any derivatives based on a specific underlying cannot be unique any more, the approach in this kind of incomplete markets is to avoid arbitrage opportunities by pricing these derivatives in *consistency* to each other.

When defining and pricing derivatives it is always an (implicit or explicit) assumption that the respective underlying market is of perfect competition. That means, for the underlying itself there are no arbitrage opportunities. Although this is only a theoretical model, as Hull (2008) explains, it is still reasonable to assume it in any real market with sufficient trading volumes, as any arbitrage opportunities there would be exploited instantly, making them vanish in very short time periods and, thus, rendering them negligible. In case of Web services, this means that providers will charge the same price for equivalent services, as otherwise they would be driven out of the market by their competitors or realize unnecessary losses. As for now there are only regarded requesters that intend to actually use the reserved service, they do not play a major role for the perfect competition in the market (that is, by reselling). Traders on the market that show this kind of behavior are discussed in Sect. 8.2.4. Note that this does not necessarily hold for unique or specialized Web services offered only in very limited quantities. In this case, derivative models are not applicable any more, and the appropriate way to implement reservation models is through yield management methods, see Sect. 8.3.

Therefore, in the following model, which can be found in Björk (2009), each Web service is offered to the requesters by any market mechanism[3] at the same price. As it is common in the derivatives literature, this price is then given by a stochastic process X that can be stated as

$$\mathrm{d}X(t) = \mu(t, X(t))\,\mathrm{d}t + \sigma(t, X(t))\,\mathrm{d}W^P(t). \tag{8.2}$$

[3]There is no need to specific this market mechanism in any more detail as only the price is relevant for the derivative model.

In this formulation, μ is a *local deterministic drift*, which is generally linked to the long term evolution of the price (i.e., some kind of trend), and σ the *diffusion term* for the scalar *Wiener process* W^P under the objective probability measure P and time t (i.e. the volatility or degree of fluctuation of the process). In addition to this, in order for the model to work, there must exist a risk free asset, which is usually assumed to be some kind of bank account or similar investment opportunity not bearing any risk. The dynamics of this asset are given by

$$\mathrm{d}B(t) = rB(t)\,\mathrm{d}t. \tag{8.3}$$

Any amount invested in this asset increases steadily in respect to the (in this case constant) interest rate r.

A derivative that has a fixed date of maturity (also: expiration date) T, and that depends only on the price at this time, that is, $X(T)$, is called a simple contingent T-claim. Let $\mathscr{Y}$ be such a claim. Then the value of $\mathscr{Y}$ is given by $\mathscr{Y} = \Phi(X(T))$, where Φ denotes the contract function. Due to the absence of any directly resellable underlying, and, thus, the incapability to replicate the behavior of the contract function of $\mathscr{Y}$, the price for acquiring this claim at $t_0 < T$ is not unique. This situation cannot be changed by any means. However, what can be done is to price any other contingent claim that is additionally defined on the very same underlying in accordance to $\mathscr{Y}$. Let the new derivative be defined by $\mathscr{Z} = \Gamma(X(T'))$, that is, $\mathscr{Z}$ depends on the same underlying stochastic process X, but has a different contract function Γ and an arbitrary exercise date T'. Now, $\mathscr{Z}$ must be priced in consistency to $\mathscr{Y}$ in order not to introduce any arbitrage opportunities which might arise otherwise by trading both derivatives simultaneously. For every time spot $t < \min(T, T')$ let $F(t, X(t))$ and $G(t, X(t))$ be the market price processes of $\mathscr{Y}$ and $\mathscr{Z}$, respectively. Both processes, though their actual value is currently unknown, depend on the same underlying price process X, and, therefore, on the same only source of uncertainty, that is, the Wiener process W^P. This allows us to follow the idea of the Black–Scholes approach, that is, to form a riskless portfolio consisting of the risk free asset and the two derivatives. In this way, the underlying, which itself formed part of the portfolio in the Black-Scholes model (or analog: any pricing formula for arbitrary derivatives) is substituted by the second derivative. This means that, while in the unique derivative pricing model consistency was always achieved with respect to the underlying, it is now in respect of derivatives themselves, it is also clear that since the underlying is not directly tradeable, there cannot arise any arbitrage opportunities between mispriced derivatives and their related Web services, but only between any derivatives themselves. As any monetary means may be invested either into the riskless portfolio or alternatively into the risk free asset, it is clear that the portfolio itself must also have a return rate of the interest rate r.

The following formulas give only a basic notion of the steps involved in the derivative pricing model in incomplete markets. For further details please refer to Björk (2009). Applying the *Itô formula* onto the price process F of the first derivative $\mathscr{Y}$, this process is equivalently determined by

$$\mathrm{d}F = \alpha_F F\,\mathrm{d}t + \sigma_F F\,\mathrm{d}W^P, \tag{8.4}$$

with

$$\alpha_F := \frac{\dfrac{\partial F}{\partial t} + \mu \dfrac{\partial F}{\partial x} + \dfrac{1}{2}\sigma^2 \dfrac{\partial^2 F}{\partial x^2}}{F}, \tag{8.5}$$

and

$$\sigma_F := \frac{\sigma \dfrac{\partial F}{\partial x}}{F}. \tag{8.6}$$

In this short notation there is $x = X(t)$ and the arguments $(t, X(t))$ for the functions F, μ, σ, α_F and σ_F were suppressed for readability. The functions dG, α_G and σ_G are defined analogously. In order to make the portfolio riskless the source of uncertainty, that is, the impact of the Wiener process W^P on the portfolio must be removed by assigning certain weighted amounts of each derivative to it. However, the in this way formed portfolio is only locally riskless, which means it stays riskless only for a very short period of time until the parts of each derivative and the risk free asset need to be changed in accordance to the continuing realization of the price process X. As being riskless, in each of these (theoretically infinitesimally small) time periods the portfolio must return the risk free interest rate r. Straightforward algebraic operations eventually yield

$$\frac{\alpha_F(t) - r}{\sigma_F(t)} = \frac{\alpha_G(t) - r}{\sigma_G(t)} =: \lambda(t). \tag{8.7}$$

In this formulation, each side of the equation depends on one derivative only and an additional parameter, the *market price of risk* λ. It can be shown that this market price of risk is the same for all derivatives defined on any underlying with the same price process X.[4] The advantage of this model and its formulation is that all other uncertainty factors, like market participants' varying risk attitudes or liquidity preferences, which might be hard to measure quantitatively, do not play any further role. The result is one aggregated risk measure λ for the price process X and any other (probably unknown) factors related to this process. In case $\lambda(t)$ is known for all t, the price of any derivative can be uniquely determined (with respect to the consistency to the other derivatives). In order to calculate this price, one can either solve a boundary valuation problem (which is not illustrated in detail here) or by using the (more intuitive) *Feynman-Kač representation*

$$G(t, X(t)) = e^{-r(T'-t)} E^Q_{t, X(t)}[\Gamma(X(T'))], \tag{8.8}$$

with the dynamics of X under Q given by

$$dX(t) = \{\mu(t, X(t)) - \lambda(t, X(t))\sigma(t, X(t))\}\, dt + \sigma(t, X(t))\, dW^Q(t). \tag{8.9}$$

[4] This result is also already known from the Capital Asset Pricing Model (CAPM) theory, see, for example, Sharpe (1964) and Ross (1973).

The approach ensures that the second derivative $\mathscr{X}$ is priced in a risk neutral valuation under the martingale measure Q (see Harrison and Kreps (1979)). Following this procedure, any additional derivative defined on the same underlying can be priced accordingly in order to be in consistence with the ones already on the market. In this way, the efficiency of the market is preserved. As can be seen, in an only internally consistent market the consistent pricing approach depends (in analogy to derivative pricing in complete markets) just on the martingale measure Q (or, equivalently, on λ itself, dependent whether one chooses to calculate the derivatives price using (8.8) or by solving the boundary value problem).

Above model contains a hidden aspect which is not readily evident. While the stochastic price process X under the measure P can be observed empirically (and, thus, μ and σ can be determined as well), matters are different for the market price of risk λ. This results from the model's assumption that the first derivative $\mathscr{Y}$ is already actively traded in the market, even though its price cannot be determined uniquely (which is also not necessary from the viewpoint of avoiding any arbitrage possibilities). Therefore, in order to determine the martingale measure Q that is essential for the consistent pricing, the market price of risk must be observed prior to this on an already active market. One method how to achieve this, as pointed out by Björk (2009), is to start with a parameterized family of functions $\lambda(t, X(t), \beta), \beta \in R^k$ and a certain amount of prices at which the derivative $\mathscr{Y}$ was traded, denoted by $\widetilde{F}_i(t, X(t)), i = 1, \ldots, n$. Then, the theoretical pricing functions $F_i(t, X(t), \beta)$ are computed and the least squares minimization problem for $t = t_0$ and the known price $X(t_0)$ is solved by

$$\min_{\beta \in R^k} \left[\sum_{i=1}^{n} \left\{ F_i(t_0, X(t_0), \beta) - \widetilde{F}_i(t_0, X(t_0)) \right\}^2 \right]. \tag{8.10}$$

In this way, the optimal parameter vector β is determined by the minimization of the error between the two prices.

In the model depicted above, the stochastic price process X was assumed to be dependent on one single source of uncertainty only, that is, the Wiener process W^P, see (8.2). In case of the trend μ and variance σ being constant over time, this process is a geometric Brownian motion. This model is widely used and accepted in traditional financial literature to model stock prices. However, with the recent advent of the availability of financial high-frequency data (see, e.g., Engle (2000)) and the spreading of derivative trades to other markets, it has become evident that this price process model is no longer neither adequate nor accurate in describing the dynamics of the underlying asset. For example, the prices of many physical commodities are known to exhibit seasonalities, while the prices for electricity contracts involve jumps of several kinds (see Seifert and Uhrig-Homburg (2007)). In these advanced and more realistic models, price processes consist of a deterministic and a stochastic part. The deterministic component is typically subdivided into a long-term trend and regular seasonal patterns, while the stochastic component contains several stochastic processes of its own, each describing different aspects, like short-term fluctuations, jumps, and spikes. With these models it is possible to capture the behavior of most today's empirically observed price models, that have to account for specific features (additionally to the ones named above), like mean reversion and regime changes. With Web services becoming a commodity of their own it is only reasonable to assume that their price processes will be subject to similar characteristics and that several deterministic components and stochastic processes will be necessary to model their behavior with sufficient accuracy.

Björk (2009) points out that the extension to multiple dimensions is straightforward. For now, in this case existing several sources of uncertainty (e.g., several Wiener, Poisson or other stochastic processes) the model generalizes to a market price of risk vector and a volatility matrix. The solution of finding a unique consistent price for any derivative is completely analog to the above model, with one crucial difference. In order to eliminate all $m \geq 1$ sources of uncertainty, it is necessary to also fix m different benchmark derivatives (defined on the same underlying) before any consistent pricing can be achieved. This means in return that, if either a too complex model is set up for the price process or there are too few traded derivatives based on this process, no consistent pricing can be done. Therefore, if one aims at setting up consistent derivative pricing in real Web service markets, one has to pay attention to the quantity of different kinds of derivatives traded on the market for a particular Web service, and, where appropriate, restrain the complexity of the Web service's stochastic price process model.

8.2.4 Impact, Limitations and Other Aspects of Web Service Derivative Markets

With the recent development in Web service commerce and the steadily increasing trading volume of these electronic services it is commonly recognized that these markets already play a vital role in today's IT and related sectors and will continue to do so in the future, also due to the advances in electronic trading platforms (see, for example, Weinhardt et al. (2006)). As pointed out in Chap. 5, in case these markets are not regulated, they will tend to shift from fixed to flexible pricing models. This is specifically the case when the markets are easily accessible, resulting in highly dynamic supply and demand changes. In today's Web service markets, there are still considerable limitations regarding the standardization of these services and universal accessibility and, therefore, no actual statement about their future development can be made with certainty. However, a comparison to commodity markets trading all kinds of tangible goods or markets enabling the trade of service-like intangible goods (like electricity or Google's Adwords auction, see Edelman et al. (2007)) admit the almost obvious supposition that Web service markets, once more established, might undergo a similar development. Since a substantial part of the traded Web services are directly related to business and industrial processing purposes, in consideration of the already today high trading volumes, the aspect of risk management will become as crucial and naturally to Web service market participants as in the above mentioned other markets.

In all above cited markets and exchanges that assumed flexible and dynamic price models, derivatives have become the preferred choice to hedge against these risks of unavailability and uncertain prices (marking the aspect of reservation and cost accounting, respectively). These risks are even fostered by the fact that the amount and variety of traded Web services are still sparse, due to specific requirements from the requester's side or specialized features the provider offers. Particularly for business process-critical Web services a non-availability of these services may lead to severe consequences in any subsequent steps, which, however, are challenging to model in general. Therefore, for the reader's convenience, the above model and the one in the following section are restricted to the standard view in these areas, that is, only the direct monetary aspect is regarded.

In case derivatives are eventually used to handle those risks the benefits apply to Web service consumers as well as providers. Besides the ability to hedge against the above named kind of risks, positive secondary effects are as follows. Although in above model only equivalent Web services were considered, in reality this scenario may not hold true, rendering sudden process adaption costs in case the Web service is acquired on the spot market directly. The possibility to use derivatives defined on a set of such standardized services enables the consumer to diminish these costs, as it makes long-term planning more feasible. A long-term planning horizon, realized through a more accurate prediction of the amount of incoming Web service requests, has also a favorable impact on the business of the provider. The requested workload can be determined with higher precision and the required resources can be adapted in the long run, with their operational costs lowered accordingly.

As pointed out above, one of the most important requirements when using above pricing model for derivatives in incomplete Web service markets is that the market for any derivative, which takes part in any pricing portfolio, must be liquid. That is, each of these derivatives must be tradeable without any delay at the actual market price. While this particular requirement must hold, on the other side one can relax the assumption of a liquid market for the Web services themselves, as the portfolios used to determine the price do not depend on them anymore. Therefore, in these incomplete derivatives markets, the liquidity of the derivative and its respective underlying are detached from one another. However, this holds only in theory. In practice, since usually the number of available Web services is limited, the number of reservations on these services will be limited as well. Thus, in case a service is of a very specialized kind and only offered at very low amounts, one cannot expect a liquid derivative market for this underlying. Though above pricing formulas can be applied nevertheless, in these cases the prices determined by a portfolio containing one or more illiquid derivatives, will not be fair, that is, not correct we respect to the derivative's real (theoretical) value, enabling undesired arbitrage opportunities.

There exist two solutions to cope with the above pointed out challenge of limited tradeable Web service derivatives. Both solutions aim at increasing the total number of derivatives on the market, and may also be combined. The first approach would be to diminish the amount of different derivatives for Web services that essentially provide equivalent functionality and quality of service. This can be achieved by enhancing the standardization of Web services between different providers and, thus, make Web service derivatives applicable on any equivalent underlying Web service independent from its specific provider. However, in this case, if the customer is indifferent from which provider he eventually will acquire the service, a fair distribution of reservations onto these providers needs to be assured. Note that the price of such standardized Web services will follow the same price process according to the rules of markets with perfect competition. The second approach would involve to admit derivatives on the market that are not traded to hold reservations but for the purpose of investment only. This allows traders to participate in the market even if they have no interest in using any Web services. Similar to financial, commodity and electricity markets these derivatives can then be settled in cash at the time they are contracted for. In this way, the amount of Web service derivatives traded on the market is theoretically unlimited and can be detached from the number of offered services. In addition to these free traders it is also possible (and in many markets indeed common practice) to install institutionalized market makers, acting on behalf of the Web service derivative exchange itself to provide additional liquidity to the market.

8.3 Yield Management Techniques for Web Service Providers

In the previous section, there was considered a market with perfect competition, that is, equivalent Web services of different providers were interchangeable and thus, due to the force of the market itself, offered at one price only, with consumers assumed to be having no preferences from which provider they acquire the services. However, although from today's perspective (technological as well as organizational) this kind of markets may very well be realized in the near future, as pointed out earlier in this book, this is not what can be observed in actual markets. Today's situation persists of several independent Web service providers with the access to their services only, available through unique interfaces which makes them hardly interchangeable. Thus, due to the lack of standards, consumers will usually face high migration costs, caused either by technological or organizational incompatibilities. Also, many Web service providers offer specialized services for which there is no substitution. These circumstances render Web service markets with the lack of perfect competition. This yields additional market powers to each Web service provider that use these to set their own prices, mainly independent of other Web service providers paying them little to no consideration at all.

In this kind of markets, which very well reflect the situation one finds today, there is a predominant imperfect market competition, that is either a monopoly or oligopoly for specific Web services. Under these circumstances, Web service providers can set their own prices accordingly in order to optimize (that is, maximize) their (expected) revenue. This is usually done with yield management approaches,[5] that is, certain models that build upon historical data and assumptions about current and future behavior and development in Web service markets.

These kind of models have already established themselves with great success, with the most prominent examples in the airline and hotel reservation industry (see, e.g., Belobaba (1987) and Vinod (2004), respectively). Common among all these application areas is that the good to be sold is a perishable one, that is, after a certain time spot has passed, it can not be used any more to generate further revenue. This applies to unused airship seats, empty hotel rooms as well as unused Web service capacities. This inefficient utilization denotes a waste of resources, yielding less revenue, also as the infrastructure for providing these goods still has to be maintained and can be adapted accurately only in very specific cases (e.g., by shutting down server capacities). By setting the prices accordingly, Web service providers can now use these in other industries already well known approaches as an additional tool to optimize their day-to-day business. Lowering the price will usually attract new customers (especially, if by this measure their current and migration costs are underpriced), while on the other hand raising the Web service offers must be done without frightening off actual consumers. Eventually, this means that the revenue will be maximized by identifying the right price for the right service at the right time.

Finally, note that the field of yield management research knows two main branches: Inventory control and dynamic pricing. While the latter was already depicted above (that is, by setting adaptively the price depending on (expectations about) offer and demand), the former deals with a slightly different scenario: Also, the provider have only a limited amount of resources available to offer their services. However, contrary to dynamic pricing, consumers approach the provider with an offer on their own, that is how much they are willing to pay for the service provided now or in the future. In turn, the provider now only has to decide whether to accept or deny their request, thus optimizing his revenue by managing his available capacities. Although this also constitutes an interesting perspective for future Web service yield management models, since this is not consistent with current most successful developments in other industrial areas, this perspective is not considered here, with the focus instead on the aspect of optimal dynamic pricing.

8.3.1 Optimal Dynamic Reservation Pricing

The model proposed by Meinl et al. (2010) for optimized Web service pricing in the context of yield management is the basic model presented by Zhao and Zheng (2000), which is also the basis for succeeding papers and advanced models (see Elmaghraby and Keskinocak (2003) for an overview). In accordance to their model,

[5]The term revenue management is omitted here. Though some authors see slight differences between the two terms of yield and revenue management, this section follows the general notion of most academic papers treating them equally.

Web service providers have always only a certain limited number of available services they can offer at each time spot. Each service is delivered instantly, that is, any services requiring several consecutive time spots are not considered. Though this advanced model is well known in practice (e.g., it is very common with hotel reservations) in this section only the basic model is discussed. Thus, these consecutive services, like storage services, have to be requested separately for each time spot. As a consequence, one can also assume the available capacity the Web service provider offers for each spot to be independent from another.

The model most employed by Zhao and Zheng (2000) is a continuous time model, that is, the Web service requesters arrive continuously in time according to a non-homogeneous Poisson process $A_{\lambda(\cdot)}(t)$ with $\lambda(t)$ denoting the process' intensity. This means that there is at most one service reservation request at each time spot, as in concordance to the model it is assumed that each customer has one request only. One notes that, since the intensity for distinct reservation time spots t_{r_1} and t_{r_2} may be different, in practical applications one would have to deal with one intensity for each reservation spot, that is several $\lambda_{t_r}(t)$. Due to the timely independence of these spots (regarding reservation and capacity), this model deals with one t_r only. Hence, this identifier may be omitted, as well as reservations for consecutive time spots.

When a customer arrives at t_c and requests a Web service for $t_r > t_c$, the task of the Web service provider is now to optimize its expected revenue or yield by setting the reservation price R_{t_r}. In the basic model of Zhao and Zheng (2000), which serves also as a basis for many advanced approaches, a closed form solution is available, though it depends not on any other providers.

The customer, after having learned this price, may either accept or refuse the offer. The latter decision can be interpreted that the customer thinks that the offered reservation price is too high and he will attempt to acquire the service on the spot market itself. In the actual model, this decision is modeled as a random variable α, which may depend on the time left until t_r and the offered R_{t_r}, that is, $\alpha(t, t_r, R_{t_r})$, denoted by $\alpha(\cdot)$, for short. Combined with the Poisson arrival process, the *demand process* for service resources available for reservation at t_r is given by $A_{\lambda(\cdot)\alpha(\cdot)}(t)$.

The random acceptance variable $\alpha(\cdot)$ can be seen as the equivalent to the *reservation price distribution* of the model in Zhao and Zheng (2000). However, reservation price in their context denotes the maximal price for which the customer accepts the offer, and may be confused with the actual use of the term *reservation*, which is why this notion is avoided here. In Zhao and Zheng (2000) each arriving customer has a fixed reservation price assigned to him according to the reservation price distribution which is assumed to be known to the Web service provider. Thus, the customer acts in a deterministic way once R_{t_r} is known. This assumption is relaxed and another point of view is adopted in the way that the actual model admits irrational behavior amongst the customers. Thus, their decisions are modeled as random variables which are then observed by the Web service provider and estimated, yielding an according distribution. Of course, from the perception of modeling eventually both approaches are equivalent, though the motivation and derivation behind these are different.

8.3.2 A Model for Returning Customers

The in the last section presented model, though extended by several authors, still lacks one important factor which up to date has not been considered in the yield management literature. This factor is the aspect of returning customers, that is, customers that arrived at t_c, rejected the provider's offer, but may return with a certain possibility later to make the same request again, in hope for a better offer. In the model of Zhao and Zheng (2000) the customers, once they rejected an offer, do not return, thus, they are out of the system and the model. Although technically feasible via logging techniques in the airline sector or hotel industry, no model in neither of these fields considers this aspect, though it certainly reduces the accuracy of these models. In the following, the ideas presented by Meinl et al. (2010) are outlined.

In the previous section, the customer arrival rate was modeled as is common by a non-homogeneous Poisson arrival process with time varying intensity $\lambda(t)$. While this intensity is allowed to dynamically evolve over time and though it is generally accepted that this represents real scenarios in many application areas (e.g., airlines, hotels, and restaurants), in most analytical research models this intensity is set as a constant (see Elmaghraby and Keskinocak (2003)). In practice, this may be due to difficulties of either measuring or predicting $\lambda(t)$.

This idea can be further extended by allowing that the intensity does not only change deterministically over time, but is a stochastic process itself. The resulting Poisson processes are also known as a Cox processes (see Snyder (1975)). For reservation requests in the Web service area it can be assumed that, as requesters need to state their exact needs on the service, the Web service provider is capable to distinguish between customers making a request for the first time or repeatedly.[6] One observes in practice that this is a common behavior among service requesters, as they tend to poll the reservation price repeatedly over time in hope for a better offer, until they either accept the offered price or leave the system. At this point, it is not assumed that the Web service provider collects detailed information about its customers. Implicitly, this model assumes that all customers behave identically.

Measuring the customer arrival rate that make a request for the first time naturally yields $\lambda(t)$ and serves as the prediction for future time spots. From the customers that return one can derive the probability that the customers, which once or repeatedly rejected an offer, will return at a later time spot to make the same request. In the actual model this probability depends on the remaining time until the spot t_c for which the reservation was requested at t_r, that is, $t_r - t_c$, and the price R_{t_r} offered this time. Thus, the customer return probability is denoted by $\rho(t_c, t_r, R_{t_r})$, or $\rho(\cdot)$ for short.

[6]This is in contrast to many today's reservation systems, for example, in the airline industry, where it is still possible to make these kind of requests completely anonymous.

The new stochastic intensity $\widetilde{\lambda}(t)$ can now be stated as follows. It is assumed, w. l. o. g. that there are no reservation requests for t_r before t_c, thus, the very first customer arrives at t_c. Therefore, the intensity $\widetilde{\lambda}(t)$ equals the deterministic one, that is, $\widetilde{\lambda}(t) = \lambda(t)$ for all $t \leq t_c$. From this point, the intensity becomes stochastic, as one does not know whether the customer accepts the offer or not, and, in case of the latter, whether he will return. This is modeled by the new intensity $\widetilde{\lambda}(t)$, $t_c < t < t_r$, which must be larger than $\lambda(t)$ due to the increased probability of one additional customer (i.e. the returning one) to arrive. As the expected amount of arriving customers for the original Poisson process over an interval is calculated according to $\int_{t_c}^{t_r} \lambda(t)\,\mathrm{d}t$, this yields the equation

$$\int_{t_c}^{t_r} \widetilde{\lambda}(t)\,\mathrm{d}t = \int_{t_c}^{t_r} \lambda(t)\,\mathrm{d}t + \rho(\cdot). \tag{8.11}$$

Unfortunately, no further statements about the exact structure of $\widetilde{\lambda}(t)$ can be made. This is due to the fact that returning customers might return with respect to different models or assumptions. For instance, if one assumes that any customer returns until t_c according to a uniform distribution, one would have

$$\widetilde{\lambda}(t) = \lambda(t) + \frac{\rho(\cdot)}{t_r - t_c} \tag{8.12}$$

for $t_c < t < t_r$. If, however, the model assumes that the customer only returns the latest until a certain time spot, for example, $t_c + (t_r - t_c)/2$, or not at all, this would be modeled by setting

$$\widetilde{\lambda}(t) = \lambda(t) + \frac{\rho(\cdot)}{\frac{t_r - t_c}{2}} \tag{8.13}$$

for $t_c < t \leq t_c + \frac{t_r - t_c}{2}$, and $\widetilde{\lambda}(t) = \lambda(t)$ for $t_c + \frac{t_r - t_c}{2} < t < t_r$. Additionally, one also has to account for customers probably arriving according to a different distribution. All this has to be reflected in the right-hand side in (8.12), which reflects the model for the customers' return behavior, while (8.11) always holds. Thus, each Web service provider's choice of the stochastic intensity model depends on his assumptions about his customers, derived from his own (or related Web service provider's) historical data.

For technical reasons, as at this point $\widetilde{\lambda}(t)$ cannot be specified explicitly, the condition

$$\int_{t^*}^{t_r} \widetilde{\lambda}(t)\,\mathrm{d}t < \infty, \tag{8.14}$$

can be stated, which must hold for all $t^* \in [t_c, t_r]$. This is a necessary condition stated in many models containing non-homogeneous Poisson processes (either deterministic or stochastic). It ensures that the amount of requests can not become infinite.

At this point two technical subtleties must be noted which at first might not be obvious. First, $\rho(\cdot)$ denoted the probability that a customer returns, that is,

$\rho(\cdot) \in [0, 1]$. However, as this notion is based on a stochastic arrival process, no return can be guaranteed, as it is only possible in this case to speak of expected return, that is, the expected amount of incoming requests has risen by $\rho(\cdot)$, while there is no guarantee that the actual occurring events generated by the process will actually rise by that amount (though they do it in expectancy). Therefore, the probability of return holds only in expectancy with respect to the total amount of arriving customers. A guaranteed return cannot be modeled by a fixed $\widetilde{\lambda}(t)$, as it would mean that one additionally would have to continually adapt the stochastic intensity as long as the (specific) customer has not returned yet. Second, it is finally noted that in some parts of the available literature, the stochastically self-adapting intensity $\widetilde{\lambda}(t)$, would rather be called a *conditionally deterministic* process, see for example, Bauwens and Hautsch (2006). Their notion is that this term is appropriate, since the intensity function $\widetilde{\lambda}(t)$ just depends deterministically on the observable history $\mathscr{F}_t$ of its own stochastic process $A_{\widetilde{\lambda}(\cdot)}$. In other words, one can interpret $\widetilde{\lambda}(\cdot)$ as a deterministic function $\widetilde{\lambda}(t, \mathscr{F}_t)$. Only in case the intensity process $\widetilde{\lambda}(t)$ would depend on an *unobservable* history $\mathscr{F}_t^*$ (which in return would also be modeled as another stochastic impact factor on $\widetilde{\lambda}(t, \mathscr{F}_t)$), it would be called a *doubly stochastic Poisson process*, that is, a Cox process.

The question that now arises is whether the stochastic intensity $\widetilde{\lambda}(t)$ needs to be adapted in case a returning customer eventually accepted the new offer at $t^* > t_c$. This can be argued from two point of views. First, considering common Poisson processes, their intensity is never adapted once an event occurs. However, in this case the intensity is changed based on the expected amount of customers. That means, if one of the returning customers finally accepts the offer, one might argue that the intensity should be changed accordingly to reflect this new information (about the lessened amount of expected arriving customers) and system state. However, this means in return that for all customers that accept an offer, the intensity has to be modified, too, if the initial intensity $\lambda(t)$ was also built up or took into account the information about the expected total amount of incoming requests. At this point it is still unclear what procedure will lead to better (that is, more accurate) results. Further empirical analyses, once these kinds of reservation systems for Web services have established themselves, are needed in order to confirm whether or not this aspect makes any difference in practice.

After adjusting $\widetilde{\lambda}(t)$ for the first arriving customer, this has to be done each time a customer (returning or for the first time) arrives, that is, each time $A_{\widetilde{\lambda}(\cdot)}(t)$ generates a new event. Thus, at each $t_c < t^* < t_r$ the process will or needs to be adapted with the conditional probability

$$\{1 - \alpha(\cdot) \mid A_{\widetilde{\lambda}(\cdot)}(t^*) > A_{\widetilde{\lambda}(\cdot)}(t)\} \tag{8.15}$$

for $t_c < t < t^*$. Of course, unlike in (8.11), this new adapted process has to be based on $\widetilde{\lambda}(t)$ itself.

It is concluded that for the case of returning customers the arrival process $A_{\widetilde{\lambda}(\cdot)}(t)$ is modeled as a non-homogeneous Poisson process with stochastic intensity

$\widetilde{\lambda}(t)$. Additionally, this intensity's stochastic behavior depends on the realization of the Poisson process $A_{\widetilde{\lambda}(\cdot)}(t)$ itself, whose characteristic in turn is defined by the very same $\widetilde{\lambda}(t)$. Thus, the stochastic process and its characterizing intensity are determined by the process' own realization and are directly dependent on one another. It is clear that generally, except for particular models, there cannot be given any closed form solution to calculate the expected number of arriving customers and their requests. This is given to the fact that the essential parameters $\alpha(\cdot)$, and $\rho(\cdot)$ depend on time as well as on R_{t_r}, which is the price yet to be optimized, in turn also dependent on future events again realized through $\alpha(\cdot)$, and $\rho(\cdot)$. This dual dependency will on most cases be impossible to handle analytically, which is why it is suggested to maximize the Web service provider's revenue using simulation techniques.

The simulation will be realized as follows. At each t_c, when a request comes in, the simulation for deriving and approximating the optimal price is run, simulating any further events until t_r. This is usually done by using Monte Carlo simulations or similar techniques. Future events at $t_c < t^* < t_r$ consist of simulated incoming requests (of both customer types, that is, returning and first-time). For each such new simulated request, the provider has to start a new Monte Carlo simulation, in order calculate the simulated optimal price the Web service provider would offer in the future at t^*. This procedure yields a structure of nested Monte Carlo simulations. The question that arises is whether this simulation is still computationally tractable.

By assuming $\rho(\cdot)$ and the acceptance rate $\alpha(\cdot)$ to be constant in all variables or by taking $\max \rho(\cdot)$ and $\min \alpha(\cdot)$ it is possible to derive an expected upper bound of necessary simulations: Given $E(A_{\lambda(\cdot)})$ to be the expected amount of customers arriving for the first time, then the expected amount of customers returning once is given by $E(A_{\lambda(\cdot)})(1-\alpha(\cdot))\rho(\cdot)$. The number of customers having rejected the offer for the second time and returning is thus given by $E(A_{\lambda(\cdot)})(1-\alpha(\cdot))^2\rho(\cdot)^2$ and so on. Excluding values for $\rho(\cdot)$ and $\alpha(\cdot)$ with $(1-\alpha(\cdot))\rho(\cdot) = 1$[7] one may use the result of geometric series and state the upper bound of all accumulated requests by

$$UB(\alpha, \rho) = E\left(A_{\lambda(\cdot)}\right)\left(\sum_{k=0}^{\infty}(1-\alpha(\cdot))^k \rho(\cdot)^k\right) \tag{8.16}$$

$$= E\left(A_{\lambda(\cdot)}\right)\frac{1}{1-(1-\alpha(\cdot))\rho(\cdot)}. \tag{8.17}$$

This is consistent with the intuition that as the acceptance rate tends to one and/or the percentages of the remaining returning customers tends to zero, that is, $U(\alpha, \rho) \to E\left(A_{\lambda(\cdot)}\right)$, while on the other hand one has

$$\lim_{\alpha \to 0, \rho \to 1} UB(\alpha, \rho) = \infty. \tag{8.18}$$

[7] This holds almost surely, as for any given value of α the probability of $\rho(\cdot)$ taking a value so that the property does not hold, is zero. With both α and ρ in $[0, 1]$ this would only hold for $\alpha = 0$ and $\rho = 1$.

Assume that the sufficient number of simulation runs for a Monte Carlo simulation is given by M. Thus, due to the nested structure (for every incoming request the optimal reservation price has to be computed as well, following the same steps) there are $M^{UB(\alpha,\rho)}$ expected simulation runs (though with decreasing time spans). In order to maintain computational tractability the precalculation of the simulation for fixed intervals (dependent on the kind of Web service offered) is suggested, with all customers arriving during the respective interval being treated the same, thus, yielding only an approximative optimal solution even if it can be assumed that the outcome of the simulations are exact.

8.4 Summary

In this chapter, the usage of reservation methods in Web service markets was elaborated. It was shown that the eventual model to be used depends on the Web service market structure itself, but not on the exact Web service pricing mechanism that is employed either by Web service providers or the market itself.

Although in the area of Web services there exist not yet a market where compatible or equivalent Web services are sold at the same price, it was reasoned and justified that in this case the model of Web service derivatives is the appropriate one. This derivative market provides market participants with the possibility to hedge the risks of fluctuating market prices. Yet, it requires a liquid market, that is, a sufficient number of market participants and/or traded derivatives in order to ensure that no arbitrage possibilities arise in the market. As happened in other markets, Web service derivative markets could also attract traders only participating in the trades for pure investment purposes.

The second model is already realized in today's Web service markets, that is, proprietary providers that offer not necessarily equivalent but rather specialized services that generally are not easily substitutable. In this case, these providers may dictate their prices as they think to be appropriate. However, even though changing to another provider may be connected with switching costs for their customer, it still remains an option for them. Thus, while their supply can assumed to be fixed, demand is not assured. In analogy to other business fields, like airlines and hotels, the use of yield management techniques to maximize their expected profit was proposed. Due to enhanced logging capabilities in contrast to other fields, it was also suggested to extend actual models by the factor of considering returning customers.

Both approaches are complementary and can coexist, since the market forms they are based upon cannot be encountered for the same Web service at the same time. Although both approaches were proposed by Meinl and Blau (2009) and Meinl et al. (2010) it remains yet to be seen which market form will establish itself in the near future.

References

Bauwens L, Hautsch N (2006) Stochastic conditional intensity processes. J Financ Econometrics 4(3):450–493
Belobaba P (1987) Airline yield management: An overview of seat inventory control. Transport Sci 21(2):63–74
Björk T (2009) Arbitrage theory in continuous time. Oxford University Press, Oxford
Edelman B, Ostrovsky M, Schwarz M (2007) Internet advertising and the generalized second-price auction: Selling billions of dollars worth of keywords. Am Econ Rev 97(1):242–259
Elmaghraby W, Keskinocak P (2003) Dynamic pricing in the presence of inventory considerations: Research overview, current practices, and future directions. Manag Sci 49(10):1287–1309
Engle R (2000) The econometrics of ultra-high-frequency data. Econometrica 68(1):1–22
Harrison J, Kreps D (1979) Martingales and arbitrage in multiperiod securities markets. Department of Applied Economics, University of Cambridge, Cambridge
Hull J (2008) Options, futures, and other derivatives. Pearson Education India, New Delhi
Kleindorfer P, Wu D (2003) Integrating long- and short-term contracting via business-to-business exchanges for capital-intensive industries. Manag Sci 49(11):1597–1615
Meinl T (2008) Advance reservation of grid resources via real options. In: Proceedings of the 2008 10th IEEE Conference on E-Commerce Technology and the Fifth IEEE Conference on Enterprise Computing, E-Commerce and E-Services, IEEE Computer Society, pp. 3–10
Meinl T, Blau B (2009) Web service derivatives. In: Proceedings of the 18th international conference on World wide web. ACM, NY, pp 271–280
Meinl T, Tatsubori M, Anandasivam A (2010) Enabling cloud service reservation with derivatives and yield management. In: 12th IEEE conference on commerce and enterprise computing, e-commerce and e-services, Shanghai, pp 3–10
Papazoglou P (2008) Web services: Principles and technologies. Prentice Hall, MA
Ross S (1973) The arbitrage theory of capital asset pricing. Rodney L. White Center for Financial Research, University of Pennsylvania, The Wharton School, Philadelphia, PA
Seifert J, Uhrig-Homburg M (2007) Modelling jumps in electricity prices: Theory and empirical evidence. Rev Derivatives Res 10(1):59–85
Sharpe W (1964) Capital asset prices: A theory of market equilibrium under conditions of risk. J Financ 19(3):425–442
Snyder D (1975) Random point processes. Wiley, NJ
Vinod B (2004) Unlocking the value of revenue management in the hotel industry. J Revenue Pricing Manag 3(2):178–190
Weinhardt C, Dinther C, Grunenberg M (2006) CAME-toolsuite meet2trade-auf dem Weg zum computer aided market engineering. KIT Scientific Publishing, Germany
Zhao W, Zheng Y (2000) Optimal dynamic pricing for perishable assets with nonhomogeneous demand. Manag Sci 46(3):375–388

Chapter 9
The Vision of Web Service Markets

This explosion of large-scale e-commerce poses new computational challenges that stem from the need to understand incentives. Because individuals and organizations that own and operate networked computers and systems are autonomous, they will generally act to maximize their own self-interest – a notion that is absent from traditional algorithm design.

Feigenbaum et al. (2009)

Services become a central building block of value creation in today's society. Novel technical, economic, and organizational challenges arise from their unique nature as services' provision and consumption coincide in time (Hill 1977). Recognizing and understanding the importance of an efficient design, production, and provision of services under the presence of their special characteristics is inevitable for individuals and the society to compete in today's global economy. Especially, rapid service innovation driven by the power of modularity that is inherent in the concept of services (Baldwin and Clark 2000) embodies the success factor in service-centric environments. However, when composing distributed service activities, the question of an efficient form of coordination comes to light and turns out to be fundamental to govern distributed value creation. As Web services are living artifacts that generally exist under the ownership of different economic entities which are self-interested in nature, system-wide goals are hard to achieve as they mostly collide with individual objectives and are therefore not intrinsically pursued (Parkes 2001).

Exceeding the scope of this book and the aspects covered, upcoming trends in the context of Web services open up new challenges and fields of research that are worth mentioning in the remainder of this chapter which is based on the outlook in Blau (2009).

C. Weinhardt et al., *Business Aspects of Web Services*,
DOI 10.1007/978-3-642-22447-8_9,

9.1 Allocation and Pricing of People Web Services

Hybrid complex services that involve electronic and human activities impose new challenges from an economic and organizational perspective. So far, micro-task markets such as Amazon's Mechanical Turk[1] provide a platform to leverage the power of human intelligence – the so called *crowdsourcing* – for highly specialized tasks such as image recognition. A pool of human individuals encapsulated by well-defined interfaces can be integrated in hybrid processes. A seamless integration of human work force in automated compositions of multiple services opens up further research questions to be addressed in the future. How can people services sufficiently be described and integrated into service value networks and the coordination of value creation? The challenges that arise from the service characteristic describing the fuzzyness of input and output parameters and capabilities are partly addressed by the high degree of standardization and specified description languages (e.g., WSDL, WS-BPEL), which are common sense. Nevertheless, in the context of people services, these challenges arise anew as human work force is hardly parameterizable and the scope, capabilities and quality of the output vary widely. Thus, incorporating human activities in automated processes requires well-specified task descriptions (Kittur et al. 2008). As inputs and outputs have to be carefully described the issue of quality assurance becomes even more crucial. The question arises of how these activities can be monitored in order to compute compensation transfers and apply service level enforcement mechanisms.

9.2 Allocation and Pricing of Highly Complex Application Services

In the context of application services, a current trend towards simplification is observable that enables an agile composition of highly specialized services that expose puristic interfaces and descriptions e.g., as in RESTful architectures based on the CRUD paradigm.[2] Nevertheless, complex services consist of service components that can themselves be a utility, elementary or complex service (analogue to the recursive specification in WS-BPEL). As the granularity of service components decreases, the complexity of their interfaces and necessary descriptions grows, which implies new challenges for the mechanism. As a result of the increased interface complexity and the semantic of input and output values, the computational complexity of the algorithm that solves the respective winner determination problem augments as well. This conflicts with the requirement of computation tractability which is inevitable for a mechanism in order to be realized in online systems. Furthermore, investment costs for the customization of service offers' interfaces

[1] http://mturk.com/.

[2] CRUD stands for the persistent functions create, read, update, and delete.

fostering a higher degree of interoperability rise which results in more static and less multifaceted service value networks. More complex service descriptions and interfaces also impact the elicitation and expression of preferences for different QoS levels. Service requesters have to specify their preferences for different outcomes regarding the complex service's attributes which leads to the question of how service consumers can be supported by tools and concepts to enable the elicitation and expression of preferences for complex multidimensional QoS characteristics.

9.3 Multi-Layered Markets for Utility and Complex Web Services

Service components that are traded in e.g., the complex service auction require low level resource services (utility services) to enable their deployment and assure scalability during run-time. Focusing on the infrastructure layer, it is also reasonable to trade utility services themselves independent from mechanisms to allocate and price complex services. Nevertheless, utility services expose different characteristics and therefore impose different requirements upon suitable market mechanisms Neumann (2004). There are several market mechanisms for the trade of utility services proposed in literature (Stoesser 2009; Schnizler 2007). Combining the trade of utility and complex services as depicted in Fig. 9.1, the question arises of

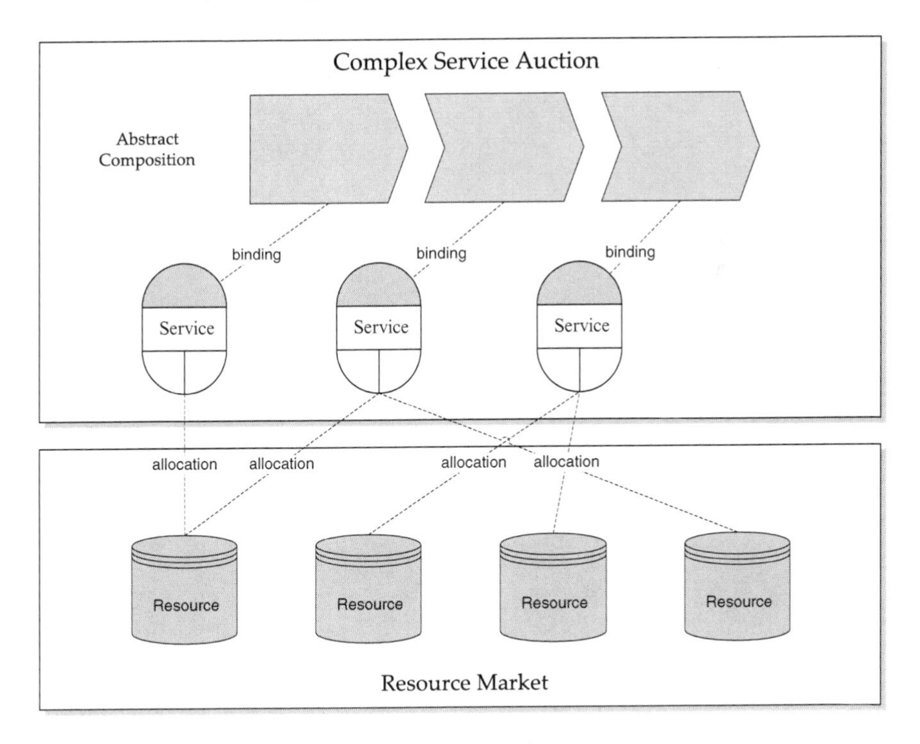

Fig. 9.1 Multi-layered market for complex services and resources

how a multi-layered market can be designed in order to enable a seamless allocation and pricing of complex services and corresponding utility service which are required by the layer above.

References

Baldwin C, Clark K (2000) Design rules: Volume 1: The power of modularity. MIT Press, Cambridge

Blau B (2009) Coordination in service value networks – a mechanism design approach. PhD thesis, Universität Karlsruhe (TH)

Feigenbaum J, Parkes DC, Pennock DM (2009) Computational challenges in E-commerce. Commun ACM 52(1):70–74

Hill T (1977) On goods and services. Rev Income Wealth 23(4):315–338

Kittur A, Chi EH, Suh B (2008) Crowdsourcing user studies with Mechanical Turk. Proceedings of the 26th Annual ACM Conference on Human Factors in Computing Systems (CHI '08); 2008 April 5-10; Florence, Italy. NY: ACM; 2008; 453–456.

Neumann DG (2004) Market engineering – a structured design process for electronic markets. PhD thesis, Universität Karlsruhe (TH)

Parkes D (2001) Iterative combinatorial auctions: Achieving economic and computational efficiency. PhD thesis, University of Pennsylvania

Schnizler B (2007) Resource allocation in the Grid. A Market Engineering Approach. PhD thesis, Universität Karlsruhe (TH)

Stoesser J (2009) Market-based scheduling in distributed computing systems. PhD thesis, Universität Karlsruhe (TH)

Index

C. Weinhardt et al., *Business Aspects of Web Services*, DOI 10.1007/978-3-642-22447-8,